DR. LAURANCE R. DOYLE

DOYLE

Reflections

of a

SETI

Scientist

PUBLISHING

Red Thistle Publications, Culver City, CA
redthistlepub@gmail.com

Edited by Steve Schwab
Layout by Helane Freeman
Cover Design by the Book Cover Whisperer - OpenBookDesign.biz
With Special Thanks to Keith Shiraki

Library of Congress Control Number 2022917235
ISBN 979-8-9868762-0-7

First Edition
10 9 8 7 6 5 4 3 2 1

Printed in the United States of America

Reflections
of a
SETI
Scientist

About the Author

Dr. Laurance R. Doyle

Dr. Laurance R. Doyle has been a Principal Investigator at the SETI Institute (Search for Extraterrestrial Intelligence) in Mountain View, CA for over 35 years, and was just the third scientist to join SETI's prestigious team. Laurance became interested in astronomy at age 6, when his Dad gave him a map of the Solar System with stars in the background, and told him the stars were other people's suns.

After receiving his Bachelor's and Master's degrees in Astronomy at San Diego State University, Laurance earned his Ph.D. in Physics in 1987 at Universität Heidelberg. Laurance has authored over 100 scientific journal articles on topics including extrasolar planet detection, animal communications and ancient African astronomy. His extensive contributions to the scientific community have put Laurance in the top 1% of cited physicists in refereed scientific literature worldwide, according to Google Scholar.

Laurance provided expert testimony to the United States Congress' Committee on Space, Science and Technology in 2013 on "Exoplanet Discoveries: Have We Found Other Earths?" He has also lectured as a Visiting Professor at Principia College in Elsah, Illinois, the University of California, Santa Cruz and the University of Paris' *Observatoire de Meudon* in France on a variety of topics, including Bioastronomy, Quantum Physics and Extrasolar Planets.

Laurance has appeared on many educational science programs, including "Through the Wormhole" with Morgan Freeman, "The Universe" with Michio Kaku and "Cosmic Odyssey" with William Shatner. He also coauthored and appeared on Richard Sergay's "Humpback Whales and the Search for Alien Intelligence," an episode of the 2021 Science and Education Webby Award-winning series, "Stories of Impact," and was a frequent contributor to space. com, where many of his articles still reside.

A triplet, Laurance currently resides in northern California, where he enjoys reading, writing and expanding his vast and diverse library. He lives by the motto, "The Universe shows you what you're supposed to be doing by making you love it!"

Table of Contents

All articles, except "Eons Play," were published on Space.com
between 2004 and 2013.

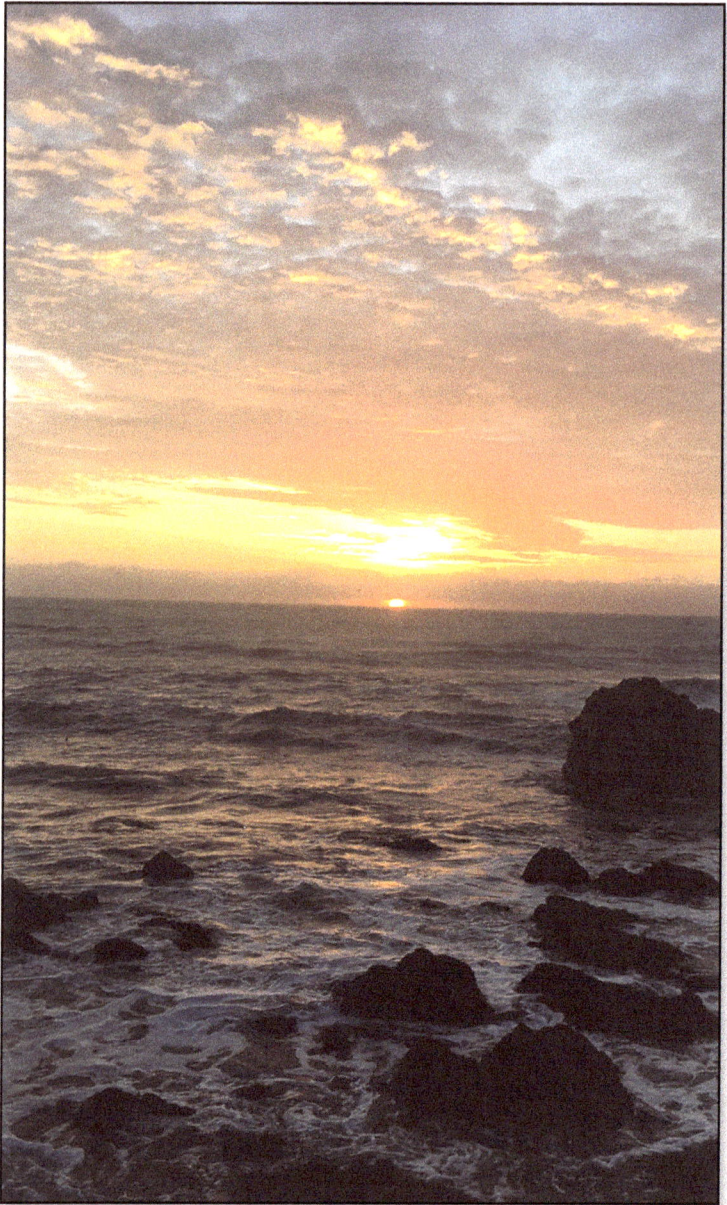

Eons Play

When the Earth was young, the Moon nearby, in a cometary sea
Prokaryotic thoughts arose, "How great it is to be!"
"Let's rust the ocean!" we agreed, "until no iron's there"
We'll use the excess oxygen and put it in the air

As huge salt mountains melted down to spice the saltless seas
The Do Si Do tectonic dance of plate activities
Stromatolites to Trilobites as O_2 filled the air
"What say we all crawl up on land and have a good time there?"

"We'll bring amphibians and trees, and oh, it will be fun!
And take some extra ozone to protect us from the Sun"
So off we went, and partied on, from Cynodont to 'saur
Time flies when one is having fun, then from a distant shore

We saw a comet hit the ground, the best I've ever seen
It turned the Moon a pretty blue, the Sun a shade of green
"Now that's a party!" we all sang, and went to mammals be
The 'saurs became a little flock of ornithology

The trees were great, but it was late, so up on two we strode
And chipped some stone and built some fires to warm the cave
abode
"Already the Holocene? My how the time does fly!
Seems like it was but yesterday the Moon was just nearby"

Now here we are, upon that Moon, next – to another sun!
A galaxy to party in, I said it would be fun!

(with apologies to Langdon Smith's Poem, *Evolution*)

A Moon by Any Other Name

Did you ever wonder how newly discovered moons and new features on planets are named?

When Galileo first aimed a telescope at the Moon in 1610, he saw mountains that looked very much like Earth's mountains. Thus, we have the lunar Alps, Caucasus and Apennine Mountains named after terrestrial mountain ranges. He saw vast, dark, lava-flooded areas that he mistook for vast seas (*mare* in Latin), and thus we have the Sea of Tranquility, where the first Apollo 11 astronauts walked on the Moon; the Sea of Serenity, the Bay of Rainbows

(*Sinus Iridium*) and the huge impact remnant known as *Ocean Procellarum* (the Ocean of Storms).

Craters and some other smaller features on our Moon are named after philosophers and scientists. For example, one of the most obvious craters on the side of the Moon that always faces us (the Moon is tidally locked to rotate at its orbital rate) is the huge, rayed crater called Copernicus. It is so large, if you were standing in it, you would not know that you were in a crater, as the horizon on the Moon is only two miles away, and Copernicus is almost 300 miles in diameter. There is a crater named Tycho, near the south lunar pole. Its rays lead to the impact crater Kepler, since Kepler, the first modern astronomer and the first to discover how the planets moved, used Tycho Brahe's observations to derive his three planetary laws of motion, so the rays seem to depict Tycho sharing his data with Kepler.

Aristarchus is another small but obvious crater; the interior is made of the brightest material on the Moon. Like Aristarchus himself, who lived in the 3rd Century B.C., and was the first to point out that the Sun was the center of the solar system, the crater, while having little impact on the surrounding area, shines more brightly than any other spot on the Moon. Socrates, Plato, Archimedes, Tsiolkovsky, Pasteur and Kuiper are all crater names on the Moon, along with others named for great ancient and modern philosophers and scientists. On the back side of the Moon, because Russia was the first to observe it (theirs was the first spacecraft to orbit the Moon), the largest maria (singular of *mare*) is the Sea of Moscow.

But what about features on, for example, the planet Mercury? These are generally named after famous authors and composers. The naming of the celestial bodies is the task of the naming committee of the International Astronomical Union; star names have recently been sold arbitrarily along with a picture of the star, but this questionable practice has never been recognized by the IAU, and such names are never used by the professional astronomical community. Thus, we have a crater on Mercury named Tolstoy, we also have the Bach crater and the Wagner mountain range there. On Venus, with one exception (the Maxwell mountains, named after the great 19th century physicist James Clerk Maxwell), all features are named for famous women. There we can find Sacagawea crater, Florence Nightingale rile, Cleopatra caldera, and the Freyja Mountains, named after the Norse woman for whom the day of the week, Friday, is also named.

As another example, on Mars, valley networks are generally named after rivers or other languages' names for Mars or planets – many ancient languages are used. For example there is Auqakuh Vallis (Incan for "Mars"), Dzigai Vallis ("valley" in Navajo), Nergal Vallis ("Mars" in Babylonian), and the Reull Vallis ("planet" in Gaelic).

Since Jupiter has no surface (none of the gas giant planets do), there are no lasting impact craters to be found there, but the innermost large moon of Jupiter, named Io, is known to be the most volcanically active moon in the solar system, with up to a dozen simultaneous volcanoes seen spouting material hundreds of kilometers into space when last seen by a spacecraft.

Features on Io are thus named after fire gods, such as Pele (the Hawaiian goddess of fire, whose work is clearly evident on the big island of Hawaii, which grows by about 30 acres each year), Loki (the trickster blacksmith of the Norse), Ho-Musubi (the Japanese god of fire) and Surt (the Icelandic volcano god). Large, ringed features on the Jovian moon Callisto, for example, are often named after mythical heavens, such as Valhalla (the heaven of the Vikings), the largest impact crater in the solar system, or Adlinda, the place deep in the ocean where souls go in the Inuit mythology.

Skipping around, we find that the largest rings of Saturn are named A, B and C for asymmetric, bright and crepe rings. Clearly, planetary scientists were not expecting the thousands of rings found there by the Voyager spacecraft! However, the dubious pleasure of having gaps (clearings in the rings) named after scientists includes Cassini (the Cassini division is a gap in between the A and B ring), Maxwell (he also figured out the nature of the rings), and Keeler, the first director of Lick observatory in San Jose, California, who observationally proved that the inner rings indeed rotated faster than the outer rings (i.e., the rings were really particles orbiting Saturn).

As other examples, many features on Saturn's moon Mimas are named after characters in King Arthur's court. There, we have the Galahad, Lancelot, Kay and Guinevere craters. The features on another moon of Saturn, Enceladus, are largely named after characters from "The Arabian Nights," and include Aladdin, Ali Baba, Sinbad and Shahrazad craters. Until the Voyager spacecraft's encounter with the Uranian system, its moons were named after Shakespearean fairies, but are now named after Shakespearean characters in general. Because of Voyager's encounter with Uranus, I had the opportunity to learn firsthand how moons are named.

I was an associate member of the Voyager imaging team during the first encounter with the Uranian planetary system in 1986. Before the encounter, five Uranian moons were known, and eight rings had been discovered by airborne astronomy. The Kuiper airborne observatory observed a star pass behind the planet Uranus to determine the depth of its atmosphere, but had found the star flashing on and off on either side of the planet. This was due to the ring material on either side.

A colleague of mine, Bruno Sicardy, of Observatoire de Paris, and I noticed that on a particular long exposure of the rings of Uranus taken by the Voyager spacecraft, one of the stars did not seem to be moving in the same direction as the others. "That's a new moon," we said, but our colleagues doubted this, indicating that it was probably one of the known moons, or that perhaps we had mistaken its motion. With the help of navigation calculations done by Steve

Synnott, part of the navigation team at Jet Propulsion Laboratory, we obtained a prediction as to when the moon, if real, would be seen in any additional pictures coming in from the Voyager imaging system. If they were real, such images would appear again going across the field of view in the pictures being transmitted the very next morning. I got caught up in traffic early the next day, but when I got in, folks on the Voyager Imaging Team greeted me with, "Congratulations! You've discovered a new moon!" Yippee! Bruno and I had discovered a whole new, tiny little world!

We called my brother, who was well acquainted with Shakespearean literature, and the decision for a name came down to "Mab" and "Peaseblossom," both fairies from Shakespeare's "A Midsummer Night's Dream." We suggested the name "Peaseblossom" then, which was well received by the scientists on the Voyager imaging team. Ah, but a controversy arose about the names of the new Uranian moons a few weeks later, when a congressman suggested that the newly discovered moons of Uranus be named after deceased astronauts – a suggestion that soon drew a reply from the then Soviet Union, pointing out that naming them after deceased cosmonauts was equally valid. This actually started a bit of an international scuffle! Because of this, no name was assigned to our little "Peaseblossom" for several years.

Finally the name "Bianca," a Shakespearean character from "Taming of the Shrew," was quietly chosen by the IAU committee on naming moons. In the Smithsonian Institution in Washington D.C., you will see that the Voyager spacecraft is credited (rightly so) with discovery of a little Uranian moon named "Bianca," about 20 kilometers across. Ah well, I still like the name "Peaseblossom," and nothing, to my knowledge, is yet named "Peaseblossom" in the heavens. However, Shakespeare is still a daily part of my life, as I have named the computer on my desk "Peaseblossom." It is there to remind me that Bruno and I were the first ever to know of a wholly new little world. And when I remember the thrill of that discovery, by any other name, it still smells as sweet.

A Spokesman for Saturn's Rings

NASA Hubble James Webb

I'd like to share how I decided on the rings of Saturn as the topic for my doctoral dissertation. I obtained my M.S. degree in Astronomy, and got a job with the Jet Propulsion Laboratory in Pasadena, California, working with the space image processing group, whose job it was to put digital images together that came back from the planetary spacecraft and enhancing them for viewing certain features, making maps, etc. We were often the first people to see new discoveries.

One of my first tasks had been to false-color Jupiter images to bring out its cloud features. Some folks in our group got the nickname

"Easter Bunny," because a lot of the time, the results frankly made Jupiter look like a psychedelic egg. I got a call from NASA Headquarters requesting a color picture of Venus. I said, "What color would you like it?" and my office mates waved at me; I had to backtrack when the voice on the other end said, "What?" since apparently our credibility for realism was on the line. However, in those days, no planet's actual color was safe in our lab.

I got a call from the boss one day; someone from the FBI wanted a ring picture enhanced. Since I had been assigned to work on Voyager I and II's Saturn ring images, he was sent to me. I didn't ask any questions as we shook hands, and he asked, "So you can enhance ring images?" I nodded, and he handed me a small 9-track tape. When I brought up the image, it was a man patting a black dog. He asked if we could zoom in on and enhance the man's hand and the ring he was wearing so they could identify him. It was no problem and he was very satisfied with the result but I found myself smiling and shaking my head through the experience.

Then, one day, Rich Terrell of the Voyager Imaging Team came in with an interesting assignment. On the processing sheet were picture numbers with the request, "Enhance for radial features." Hmm, radial features? Rich said that Jeff Cuzzi of NASA Ames had thought he noticed radial features going outward from the rings. Carl Sagan and Dave Pieri joined us at the computer screen as I loaded up the images. After a quick look at the brightness histogram, I did a contrast stretch and suddenly these massive radial features jumped out at us. "This is going to make someone a great dissertation," was the first comment from Carl Sagan. Hmm, I thought...why not me?

I began saving images of Saturn's rings that showed their different colors and aspects. Eventually, I did my dissertation in Germany with Eberhard Grün at the Max Planck Institute in Heidelberg, whose interest in these radial features – now called the "spokes" – arose from his interest in interplanetary dust. From the way light scattered off the spokes, we knew they must be small particles (i.e., near the wavelength of light). One way to determine this is that they became brighter when looked at in the direction of the Sun, but became darker when looked at with the Sun at one's back, or over Voyager's shoulder, one might say. This is exactly the same effect that happens with dust on a car windshield. One cannot see the dust very well while the Sun is at one's back, but at a certain angle, as one turns into the Sun, the dust brightens up and one sees nothing but dust. This angle, called the "phase angle," at which the "forward diffraction lobe" of the tiny particle sends light toward your eye, can actually measure the size of the particles doing the light scattering in terms of the wavelength of light, in this case, sunshine. This became the basis of my dissertation: to use multicolored images of Saturn's rings from as many Sun angles as possible to measure the size of the spoke particles.

I thought I had been there when the spokes were discovered, but research showed that as early as the late 19th century, these features had been spotted in the B-ring of Saturn. The B-ring is the middle ring, and has most of the mass of Saturn's rings in it; the mass of all the rings is about the same as Saturn's moon Mimas. In the 1930s, some folks had thought they had spotted the ring-spokes again, but by the 1960s, no one could see them and they had pretty much been relegated to the realm of illusions the eye plays, much like the canals of Mars. But what were the spokes? And how did they form so quickly (some as rapidly as ten minutes or so)? Why did they form radially outward when the ring particles move around the planet in a more or less circular orbit? All good questions, and a great subject for a dissertation.

Theories were soon being proposed, but the most promising was that the spokes were the result of meter-sized meteor impacts onto the rings. As they hit, they produced a plasma of charged particles made from the regolith that constitutes the surface of the B-ring boulders. These spoke particles were therefore charged, and while trying to rotate with the rings, were pushed into a perpendicular direction by the magnetic field of Saturn. Neat, huh? An important parameter was their size, since they would have to be big enough to "catch" electrons but small enough (i.e., light enough) to hover due to electric charge levitation. Theorists were coming up with an expected size, so I went after the observations and radiative transfer modeling that could give me the actual measurement of the size of these particles, based on the forward diffraction lobe effect mentioned earlier. There is an old saying in science that a theory is something that no one believes but the person that formulated it, and data is something that everybody believes but the person that took it. So, with apologies to the theorists, I picked a "data" dissertation.

Using Voyager images from different angles and in different colors, I finally came up with a particle size. They were just about the size of red visible light, at 0.6 microns, with a very small size distribution around this particle radius. Unknown to me at the time, the theoretical size of about 0.12 microns had been revised, and the new theoretically expected size was more like 0.6 microns! This was great too, because the theorists had not been aware of my results when they submitted their paper, and I had not been aware of their new results when I had measured this size for the particles. Great when things match up!

Soon after this, Luke Dones, Jeff Cuzzi, and I also used the images of the B-ring itself to measure the brightness of the large boulders at different Sun angles. The rings were very bright, and if they had been out in all the meteoritic "rain" since the origin of the Solar System, they should have been much darker than we measured them to be. We even tried hiding the meteoritic material underneath the surface of the boulders of the B-ring themselves, but to no avail. The conclusion we were being guided toward led directly to a very young age for the rings of Saturn, since they could not have been out in the "dark rain" for very long. We estimated something less than 100 million years, and likely even as young as 10 million years. If the dinosaurs had invented telescopes, they would likely not have seen the rings of Saturn because they would not have been formed yet. They are likely the result of a moon somehow getting too close to Saturn and breaking up due to tidal forces sometime after the Cretaceous period.

A few years after publishing these results, I got a letter from Clyde Tombaugh, the discoverer of Pluto. He said he had spotted the spokes in Saturn's rings from the ground using his 16-inch telescope. We corresponded a bit, but I had already decided that I was not going to argue with the discoverer of Pluto. But he also could pinpoint where they were, and I became scientifically convinced (and not just historically impressed) that he had indeed spotted them. A few years after this, observations done with the Hubble Space Telescope also confirmed a narrow size distribution, and everyone was looking forward to the Cassini spacecraft's arrival to get better imaging data on the spokes as well as other features.

Well, nature always has something surprising, and when Cassini got to Saturn there were no spokelike features to be seen in any of the rings! They were gone! I guess it isn't often that one's dissertation topic disappears (unless one works on supernovae or such things), but that's what had happened. It was suggested that it might be an effect of the angle of the rings toward the Sun. Or perhaps the meteor flux at Saturn had changed. But their disappearance might perhaps explain the skepticism about the spokes in the 1960s when they really could not be seen in the rings, because they actually do come and go.

Well, I wish I had a good ending of this story for you. But as of today, we don't know what happened to the spokes in Saturn's B-ring. Jeff Cuzzi and his colleagues are making predictions as to when the spokes should reappear. At their earliest prediction, the spokes did not reappear, but there are other predicted times too, based on modifications of the theory. I'd like to see them suddenly reappear someday soon.

The rings of Saturn are so beautiful already that I would expect them to become a favorite spot for future space-faring honeymooners. And if they are accompanied by the sudden-burst formations of mysterious-looking cloudy radial features shooting thousands of miles across the rings like a silk handkerchief pulled from a magician's sleeve, to then drift along softly and settle back into the majestic rings to await another burst of plasma, well, it can only add to the experience. Until then, I hope it's a matter of months or years rather than decades or centuries before they reappear. I'll await their announcement like news of an old friend returning home.

Bee Celestial Navigation and Non-Human Intelligence

Bees practice celestial navigation to communicate the position of the Sun through dance.
(Image credit: Dr. Laurance R. Doyle)

Millions of years ago, a group of wasps "decided to" become vegetarians, and so today we have the bee. Some of their cousins "decided to" quit flying and so became the ants, but that is another story. Although only about 20% of bees are social, honeybees are very social indeed. It has been stated by several biologists that if it were not for the honey bee pollinating plants, humans would only last three or four years, as our food supply would disappear.

The female honey bees are the workers of the hive. First, they learn to babysit, then they learn the construction trade (specializing, of course, in hexagonal wax structures), and eventually take on the daunting task of navigating the outside world. Honey bees have been known to travel to find honey over 10 kilometers away from their hives, which is the equivalent of a human flying from San Francisco to Denver to get some pollen. It takes the life's work of about six bees, flying thousands of miles, to make one teaspoon of honey.

Bees are the only species, other than man, currently known to communicate with symbolic language, meaning they can "talk" about the details of something that is not present. We note that psychologists dispute the use of the terms "symbolic" being applied to any non-human communication systems, but bee scientists regularly apply this term to describe bee language. And what do bees "talk" about? Mostly astronomy, or, in particular, the Sun, and where it is as compared to where the flowers are. And how do they talk? Mostly, they dance!

We know of three languages that bees use; it has been postulated that they have several more. The easiest is the Round Dance. Basically, when a bee finds a nectar source nearby, she comes back to the hive and dances around in a circle, giving out samples (for humans, this works well at ice cream stores). The Round Dance tells the other bees to go out and sniff around for the source, as it is very close.

Another dance is known as the DVAV Dance, basically a kind of bee belly dance. This dialect is reserved for internal hive politics. Who is to be the next queen? Is it a good day to swarm? And so on...

But the most studied language of the bees is the Waggle Dance. When a bee finds a nectar source farther away, she comes into the hive and gets some of the other ladies to gather around. Although it is dark, they can feel how she dances and also taste a bit of the quality of nectar she has brought back. She then starts this special dance over the combs. If more than one bee is dancing, eventually which source to go to first will be decided democratically; it is "discussed" until the vote is unanimous.

In the Waggle Dance, "up" is always the direction to the Sun. The bees have little muscles in their necks that can tell which direction is vertical in the dark. The angle from the Sun to the nectar source is then the angle at which the scout bee dances from the vertical, indicating the angle at which the others must navigate.

But how far away is the flower? As the scout bee (sometimes called the "recruiter bee") dances, the number of waggles she does in the correct angular direction before turning around to begin again is how far the honey source is, in bee units. Different types of honey bees have slightly different units of measurement. Finally, the time she takes doing the dance indicates how much of a headwind can be expected. This tells the other bees how much fuel (honey) to tank up on to make their trip there and back.

Many remarkable experiments have been done with bees over the past hundred years. How they use polarized light to see the Sun on a cloudy day; how they can understand the landscape as a map, and so don't need to follow the same route back to the hive that they took going out; how they know where the Sun is even after it sets and so can forage during a full Moon, and many more.

But I was particularly intrigued by a serendipitous experiment I read about recently that occurred when some university scientists were training bees to go farther and farther away for nectar, so they could determine the precision of their navigational directions to each other. They placed some nectar close to the hive and then moved it out 25% farther every day until, after a while, the nectar source was quite far away. This required quite precise directions from the scout bees to the others in order to allow them to find a spot this far away. In other words, the angle of the Waggle Dance had to be smaller the farther the distance away the nectar source was.

They were doing this experiment, which had been going on for many days, when the professor got a call from his graduate student. The student's car had broken down, so he had been unable to place the nectar source an extra 25% farther away that morning. The professor said he would do it that afternoon.

When the professor arrived at the nectar source, there were no bees present. But when he arrived at the place where the nectar should have been for that day, but had not yet been moved there, all the bees were there waiting for him! Not only had the bees gotten the math correct (25% farther), but the implication is that they had demonstrated the imagination to be able to picture the future by picturing the nectar "not where it was," but where it was going to be! The professor wrote that he would never have done such an experiment on purpose since he never would have thought that the bees could have been so intelligent!

Besides basically doing all the work to bring us fruits, vegetables, and other pollination-requiring plants, plus honey and beeswax - bees remind us not to underestimate the expression of intelligence from any of our fellow (or, in this case, our lady) species. So to bee or not to bee is not the question. We have to bee, and we should be grateful to have such reliable, symbiotic friends to share our planet with.

So what does all this have to do with SETI (the Search for Extra Terrestrial Intelligence)? Well, the three main requirements for producing extraterrestrial communications are a communication system, advanced tool use and astronomy. Bees demonstrate non-human skills in all three. And the more we can learn from them (and other species), the more prepared we should be for a truly alien signal if and when it is received from extraterrestrials that have not grown up on the same planet nor shared the same star with us for millions of years.

(For further details about this serendipitous experiment, see: Gould, J.L. and Gould, C.G., *The Honey Bee*, Scientific American Library Series.)

The Canary Islands Winter School on Extrasolar Planets

Early in the year 2005, over 100 graduate and post-doctoral students from 17 countries attended a special Astrophysics Winter School in the Canary Islands, located just off the northwest coast of the African continent. I was fortunate to be one of the eight invited professors. Each of us lectured on our specialty related to extrasolar planets – detection methods, observational techniques, planet formation theories, characterization of discovered planets, and so on.

The Canary Islands, belonging to Spain, are a European center for astronomy. Two of the islands, Tenerife and La Palma, are home to 31 telescopes. The location's clear night sky is well-suited to astronomy, and the astronomers usually only have to shut down due to sand that occasionally blows from the Sahara Desert. The sand gives the sky a pretty, twinkling appearance, like blue diamonds raining down from the heavens. For delicate telescope mirrors, however, the result is far from beautiful – and difficult to clean off!

During these sessions, I lectured on various extrasolar planet detection methods including Pulsar Timing, Periodic Radial Velocity Variations, Gravitational Microlensing, Astrometry, Imaging (both chronographic and nulling interferometry), Radio Flux, Transit Photometry, Phase-Reflection Variations, Eclipsing Binary Minima Timing, and, for those who believe that civilizations will transmit from planets, radio and optical SETI.

Everyone had a wonderful and educational time. Ten years ago, there were but a handful of astronomers doing research on extrasolar planets, and here was a room of over 100 advanced science students selected from around the world. With the typical Spanish agenda, including two hours for lunch, there were many opportunities for in-depth discussions. It is difficult to convey, in a few words, what an inspiring time this was for all.

After the Winter School, I traveled with my colleague, Dr. Hans Deeg, to the nearby island of Gomera, where a 2,000-year-old whistle language is still used. The people here developed this language long ago to communicate across the island's rocky terrain. We recorded some whistles from an expert there, Professor Don Mendoza, who teaches this "Gomera Silbo" language as part of each school's required curricula. We hope to be able to compare this human language of whistles directly with the whistle signals of bottlenose dolphins in our research here at SETI Institute, where we seek to understand better the Drake Equation term f_i – the fraction of life-bearing planets on which intelligent life emerges. By examining human and cetacean whistles, we can overcome the difficulties that might arise from comparing the "apples" of human phonemes with the "oranges" of dolphin whistles. Since both are whistles, we can use the same signal-classification scheme.

When the astrophysics school was visited by the Princess and Prince (and future King) of Spain, each professor was introduced in turn, having time for a little discussion. The Spanish royalty seemed decidedly interested when they read the affiliation "SETI Institute" on my nametag. They were very nice folks and well worth the effort of putting on a suit and tie for – at least for a couple of hours.

Naming New Extrasolar Planets

NASA/Hubble James Webb

There has been more public furor over the demotion of Pluto to "dwarf planet" than I have seen about other astronomical issues in a long time.

It's good to have large participation and interest in astronomy, and many folks really care significantly about the naming of celestial objects. In a recent science meeting, many colleagues discussed the naming of new extrasolar planets. I happened to disagree with the majority, and was surprised when a young scientist stated, with what appeared to be frustration (and more than a little enthusiasm), that the naming of extrasolar planets was already a tradition and could not be changed. Wow, a tradition in only 15 years! But the emotional involvement was a surprise, especially from a scientist. So here is the issue at hand.

Extrasolar planets are currently named in order of discovery, using lowercase letters of the English alphabet. For example, the first extrasolar planet-mass bodies discovered around the pulsar PSR

1257+12 are named PSR 1257+12-b, PSR1257+12-c, and PSR 1257+12-d in order of discovery ("a" is reserved for the central star). This has worked out well for this stellar system, since the planets also happen to be in order of increasing orbital period (i.e., distance from their star).

Now, we also now have Gliese 876-d, Gliese 876-c and Gliese 876-b which are in the correct order from the star, but in this case the outer planet, 876-b, was discovered first, followed by the second farthest, 876-c, followed by the closest planet to the parent star, Gliese 876-d. The star HD160691's planets (discovered so far) are, in order of distance from their star, HD160691-d, HD160691-e. HD160691-b and HD160691-c. The planetary system 55 Cancri is really fun; there, the planets, listed in order from their parent star, are 55 Cancri-e, 55 Cancri-b, 55 Cancri-c, 55 Cancri-f, and 55 Cancri-d. You get the idea. Each planetary system preserves the historic order of discovery in their names but tells us nothing astronomical about the planets or their proper relationships to each other. In addition, it seems like it could create quite a confusing situation after many planets are discovered. Perhaps it's akin to what Greek or Indonesian children must have to go through when learning the geography of their country's thousands of islands.

My suggestion, which was not adopted, was that each planet be named for its stellar parent as usual, but then be designated by its orbital period in days, to one decimal point. The orbital periods may certainly be expected to be constrained to within a tenth of a day or so. No two planets could be confused, unless there are Trojan planets which share orbits, but this may be rare, and at any rate, they might have the additional unambiguous designations of i and ii, as needed. Thus, we would have Gliese 876-1.9, Gliese 876-30.9 and Gliese 876-60.1. Those with a bent for history would have to look up the discovery order, instead of the astronomer having to sort through a list of letters to figure out which is the one to observe for a transit or for radial velocity data. Similarly, we would have 55 Cancri-2.9, 55 Cancri-14.6, 55 Cancri-44.3, 55 Cancri-260 and 55 Cancri-5218.

Now why make such a fuss about this now (if writing an article about it can be considered a fuss)? Over 340 extrasolar planets have already been discovered and things have seemed to work out OK so far, right? The reason is that current space missions may soon discover thousands of additional planets in newly discovered systems, as well as many smaller planets that we were previously unable to detect in already known planetary systems. Some space missions will be able to detect Earth- or even Mars-sized bodies. If our Solar System is typical in terms of having eight or more planets to a system, then there could be designations consisting of combinations of the letters b, c, d, e, f, g, h, and i! And all sorts of permutations on these could occur. We could get: (star name)-b, g, f, d, e, h, i, c or (star name)-e, d, i, b, h, c, f, or any other of the 46,233 possible combinations of eight letters taken one, two, three, and up to eight ways at a time. The possible combinations go up with the factorial of the number of planets, so things do not really get bad at one, two, three, four or even five planets to a system (still only 120 possible combinations, on the order of learning the states of the US). But as one gets up into six or more planets to a star system, the possible combinations really begin to take off.

And this is just the start of the fun. If we consider multiple star systems where stars are usually designated by the capital letters A, B, C, etc., we get situations where, for example, alpha Centauri A-b is a planet, while alpha Centauri B-a is a star. (The "a" designation would, of course, not be used unless a planet is discovered and then the "a" for the star is implied, using the present nomenclature.) Let's look at the six-star system Castor (alpha Geminorum), where we actually already have the stellar components named Aa, Ab, Ba, Bb, Ca and Cb. (The second components of close binaries are referred to with lower case letters in these cases.) These are each star names, but if a planet was discovered around star Ab it would, by the current nomenclature, be designated alpha Gem Ab-b, while a second planet would be alpha Gem Ab-c. Similarly, planets discovered around star Bb in order would be alpha Gem Bb-b, alpha Gem Bb-c and so on, and planets discovered around alpha

Gem Cb would be alpha Gem Cb-b, alpha Gem Cb-c, etc. So in this case, alpha Gem Bb is a star, even though the first planet around the double star 16 Cygni B is 16 Cyg B-b. I guess the hyphen would be the only way to tell. Wouldn't it be easier, if three new planets are discovered around alpha Gem Bb, to have the designation be something more like alpha Gem Bb-12.3, alpha Gem Bb-20.2 and alpha Gem-43.6? Incidentally, double star systems are not rare; more than half of all stars are in binary systems.

Again, I don't want to ruffle any fur, but with thousands of new extrasolar planets, down to terrestrial-body sizes being discoverable within the next decade or so, shouldn't we be taking stock of our existing nomenclature? I know the IAU commission on naming celestial objects is very diligent (and very busy), and so hope they have considered (along with other extrasolar planets astronomers) some of these possible permutations. You can see further interesting suggestions for nomenclature in Wikipedia. I shall leave others to do the possible combinatorics and nomenclature for future multiple-star system planet discoveries for up to eight planets around each of the stars in, say, a triple star system. The number of triple star systems is non-trivial.

I must add that this is not the first time astronomers have run into trouble with confusing nomenclature (aside from the usual beginner's confusion over larger magnitude numbers referring to fainter stars, and things like that). The historic names of variable stars is particularly amusing. A gentleman named Friedrich Argelander decided that the first variable star in a given constellation would have the designation "R" put in front of it. He chose R because he knew the earlier letters in the alphabet were already being used (for multiple star systems, for example, as noted above). But he also felt that there could certainly be no more than nine variable stars per constellation. Unfortunately, he was off by thousands and thousands. After R of course, came S, continuing up the (English) alphabet to Z. Rather than going back to A, the designation that follows Z was then dubbed RR, then RS, then RT on up to RZ. For example, I've done work on the eclipsing variable RT Andromeda.

After RZ comes SS (not SR, by the way), followed by ST, on up to SZ. Then comes TR to TZ, and so on up to ZZ. (ZZ Ceti is a famous star, for example.) OK, now what? RRR? Well, no.

It turns out that after ZZ comes AA (which, apparently, would not be confused with double stars using Aa since in this case the letters are both capitals). Then comes AB, on up to QZ finally, skipping the letter J for some unknown reason. Maybe to keep us on our toes? Since there are thousands of variable stars in each constellation (the number of which, of course, grows all the time) the 334 variable names possible with this system (up to QZ) was clearly not going to keep up. OK, so now would it be time for RRR? Alas, no. Astronomers gave up after this and began to use V 335, then V 336 and so on, with the "V" for variable star. Might have been good to think of this ahead of time but after all is it all part of the rich tapestry of the history of astronomy, right?

Now, back to extrasolar planets. Why would current astronomers prefer letter designations (not dissimilar to what early variable star folks used) rather than numbers that would mean something more astronomical? Would planetary orbital periods, for example, change? I would assert that most planets would not be changing their orbital periods by a tenth of a day or more over thousands of years, but special nomenclature might be noted (similar to the i or ii above) for such rare cases. Thus, with planetary orbital period designations we might hope to avoid impending nomenclature confusion by using numbers that also mean something astronomically (with apologies to science historians, of course).

Finally, an example of smart nomenclature, I have always felt, was the inclusion of the position of the star system in the name, like BD 16 + 516 and the pulsar system given above, PSR 1257+12. In this case the first number is the rough Right Ascension (i.e., essentially the star's longitude on the map of the sky) and the second number the rough Declination (essentially the star's latitude on the map of the sky). So, by only seeing the name it can be determined if these stars are "in season" and observable at night at this time of year, as well as if they are high enough above the horizon at a given observatory. It is true that stars change position according to the processional motion of the Earth's rotation axis, but this has not been a problem; because the designation is so coarse, precession has a rather tiny effect over a century.

Within this decade, we may expect to discover thousands of new extrasolar planets which, using the current designation, could produce tens of thousands of possible (and frankly essentially meaningless) letter combinations designating them. Therefore, I humbly make a suggestion to my fellow astronomers: it is not too late to bail on this system (as "traditional" as it is). We may not want to go through the variable star experience of the 19th century again in spite of its rich historical tapestry. But if we do, I guess I can always practice on Greek or Indonesian island names. And when I think of this as an emotional issue, what will happen if astronomers start to discover Pluto-like bodies around other stars? Are they going to be planets with letter designations or dwarf planets with other designations? I think, as they say in England, I'll give that one a miss.

PlanetQuest: Hunting for Extrasolar Planets

$$N = R_* f_p n_e f_l f_i f_c L$$

Ever want to discover a new world? That's what we intended to do with PlanetQuest, a distributed computing screen saver that would have allowed anyone to find extrasolar planets on their own computer. Like the venerable SETI@home, PlanetQuest would have allowed enabled users to discover real planets around other stars using four different detection techniques. By the way, there is also a NASA mission now named PlanetQuest. The two PlanetQuests were distinct, but shared a common educational goal to promote enthusiasm for planet detection.

The first planet detection technique that PlanetQuest planned to use was the single star transit method. Sometimes a planet's orbit will align in such a way that it moves across its parent star, creating a shadow, or transit. For a Sun-like star, the brightness drops by about 1% for about three hours if the planet is about the size of Jupiter and orbits very close to the star (with a "year" of about a week). When a planet moves in front of a single star, the light drop is periodic and fairly easy to recognize. An issue with the single star transit detection method is that stars, like lightbulbs, are brighter in the center than at the edges (and bluer – hotter – in the center as well). This is called "limb darkening." It means that when a transiting planet moves across the center of a star, the light blocked can also look like two eclipsing stars just grazing each other's outer edges. About two in three transit-looking features will be grazing eclipsing binaries rather than planet transits, so one must find out if the star being observed is a double star or not.

A specialization of the transit method is the detection of transits across eclipsing binary star systems themselves. This can be very tricky since in most cases, the planet cannot be said to be moving across the two stars as much as the two stars are moving behind the planet as they orbit each other. This produces a predictable, but non-periodic signal, that many times does not look much like a transit at all. The dimming and brightening will not be regular, and their occurrence will, generally, not repeat in the same way because the stars and transiting planet will be at different places in their orbits when the planet moves in front again. PlanetQuest's approach to this problem was to use a matching filter that compared all possible models of planet sizes and orbits to the observations. This would have required a rather huge amount of computational time, so it would have been a perfect opportunity for the public to participate with distributed computing.

The second detection technique PlanetQuest would have used was the eclipsing binary minimum timing method. This method relies on the fact that eclipsing binary stars, as they move in front of one another, are essentially a kind of "clock" that gives a time stamp to the observations. We watch the light of the stars (take pictures of them) and record the time exactly. The plot of the brightness of a star as it varies with time is called the star's "light curve." We have to correct for things such as where the Earth is in its orbit at the time of the observations, so we generally work with "heliocentric" time – the time at the center of the Sun when the eclipse occurred. This way we don't have to keep correcting for the light's travel time across the Solar System to the Earth.

This method is able to detect circumbinary planets of Jupiter-mass or larger, because these planets will offset the two eclipsing stars like a see-saw as they orbit them. The offset of the two stars due to a planet orbiting around them is detectable, because the eclipses will be early or late, depending on the direction of offset. PlanetQuesters also could have detected Jupiter-mass planets that don't even have to be orbiting across the disc of the two stars.

The third detection method is the gravitational lens planet detection. When there is a very close alignment between two stars, the star in front can bend the light from the star behind due to bending spacetime, according to general relativity. Any planets in the foreground star will also bend some light for a shorter time, and so be detectable also. Since the data for this detection method are the same as that for detecting transits and binary eclipses (wide field, crowded star field images), we would also have been able to detect planets using this method. The stars, however, brighten instead of dim. One can tell, for example, a stellar flare from a gravitational lens event because gravitational lenses are achromatic – that is, not of any color. Bending spacetime does not care what color the electromagnetic light wave is, while stellar flares, for example, are brighter in ultraviolet than in red light.

Finally, PlanetQuest would have used a new kind of SETI detector that complemented existing SETI projects. All SETI searches to date (as far as the author knows) are aimed at detecting the nature of the signal itself, rather than the content of the signal. In the case of radio SETI, a narrow-band radio carrier wave is detected, and in the case of optical SETI, a nanosecond pulse is detected. In our approach, we would classify the signals and produce a distribution of the frequency of occurrence of the signals. Then we compare this with the frequency of occurrence of known intelligent communications (this field of study is known as "information theory"). If there is a close match, we would look at the structure of the signals and their dependence on each other. An example of this would be, if one recorded babies babbling, and tried to find a connection (syntax) within the message, one would not find such structure. But we know that if we record the vocalizations of an adult human, that there are grammatical and syntax rules that are causing certain words to occur at certain times with relationship to each other. This is what allows one to fill in missing words from a copy where the copier was low on toner and parts of the text are missing. We have applied this method extensively to vocalizations of ground squirrels, squirrel monkeys, dolphins, humpback whales, and humans so far. If an extraterrestrial signal is received (if they are transmitting information), then they too will have to obey these rules of information theory.

We at PlanetQuest have completed observations at Siding Spring Observatory in Australia, and at the UC Lick Observatory in California. We called the combination of our search engine and educational materials the "PlanetQuest Collaboratory." Every PlanetQuester would have discovered something – the nature of a star, or a new planet, or many other possibilities – and received credit for that discovery in the PlanetQuest catalog, which would have been accessible online. And so, if PlanetQuest had been funded, you too could have joined the professional astronomers in discovering totally new worlds on your own!

The Lifetime of
Interstellar Civilizations

NASA Hubble/James Webb

Looking at the last term of the Drake Equation, we see that it relates to the lifetime of technological civilizations – how long they last as technological (i.e., interstellar communicating) entities. The three biggest considerations for our civilization at the moment could be characterized as a) getting along with each other; b) getting along with the environment; and c) staying technologically alert for large-scale concerns from space. As an example of the last item, the dinosaurs had over 200 million years to develop a comet deflector, but never did so. Some dinosaurs were bipedal, had opposable claws, and were pretty intelligent, so why didn't they, for example, invent space travel? Well, that's a topic for another essay. Meanwhile, let's stick to a few of the things we might want to deal with "out there" at various times in the future, from a few thousand to a few billion years from now.

Magnetic Field Reversal

The Earth's magnetic field is thought to be generated by a dynamo effect; that is, the movement of charged particles in its huge iron and nickel core as it spins. Other planets have magnetic fields also, and there seems to be some relationship between the strength of the magnetic field with the size and spin rate of the magnetic core. Jupiter, with a huge core and a 10-hour rotation period, generates a massive magnetic field, for example, and spacecraft sent there have to be specially built to withstand this intense field. You may remember the science fiction movie "Outland," with Sean Connery stranded in a mining colony on Io, one of Jupiter's moons. However, Io would be uninhabitable since the magnetic field of Jupiter causes a 5-million ampere electric current to flow through it.

However, magnetic fields can also be helpful. They can protect the inhabitants of the planet from high energy particles from the solar wind, for example. When high energy particles from the Sun encounter Earth's magnetic field, they are deflected toward the poles, causing beautiful auroral "curtains" of color as they hit the atmosphere. Without the magnetic field of the Earth, these high energy particles could do damage to biology on our planet.

When rocks containing volcanic magnetite cool, or are baked (as in clay pottery), they record the direction of the magnetic field of the Earth at the time of cooling. It turns out, from examining rocks of various ages, that the Earth has reversed its magnetic field many times, with the last reversal occurring about 750,000 years ago (the average being about every few hundred thousand years). Recent measurements of ancient pottery and other evidence suggest that the Earth's magnetic field may be declining, perhaps getting ready for an overdue reversal. This could take place within the next couple of thousand years. If the Earth's magnetic field is just beginning to reverse, it would certainly be important for us to protect ourselves from the high energy particles of the solar wind and from space. It would not be as devastating an event as, for example, a comet impact, but it does indicate that we do not have the luxury of indulging in another Dark Age over the next thousand years or so. If civilization is to maintain itself, we need to be on our technological "toes" pretty much from now on.

Moon Stabilizes Earth's Rotation

The most popular theory for the origin of the Moon is that it came from the Earth. We can calculate evolutionary histories of the Moon's orbit as it moved away from the Earth after formation. It is still moving away due to the Earth's tidal pull at the rate of about one inch per year. The majority of the tidal dragging comes from Earth's rotational slowdown, with most of it being caused by waters dragging over the fairly shallow Bering Sea. In doing some of these kinds of calculations for Mars, it was discovered that the direction of Mars' rotational axis could flip rather suddenly. Now this is not the normal "precession" (as it is called) of a few degrees that changes, for example, our north star though the millennia. Mars was calculated to have flipped its rotational axis up to 90 degrees in as little as a couple of million years. This was a result of the orbital angular momentum, under certain circumstances, being transferred to the rotational angular momentum, causing a coupling that led to such a flip in rotational axis direction.

So why has this not occurred on Earth, whose axis has seemingly not flipped by more than a few degrees? The apparent explanation is that the Moon absorbs any transfer of orbital-to-rotational angular momentum, preventing the flip.

Would such a flip be important? It could get very serious, like the time a couple of hundred million years ago, when all the continents were combined into one big continent called "Pangaea." If the Earth's rotation axis had flipped then, this one big continent would have become a polar continent like Antarctica. So, it would appear that a moon is required for a stable planet with life.

This was perhaps surprising news to folks that would like to see habitable planets widespread in the galaxy requiring, as it does, both an Earthlike planet in the circumstellar habitable zone as well as a fairly large satellite. This would seem to rule out habitable planets being very common. However, additional research into the rotational histories of the planets shows that Earth used to spin a lot faster. If the Earth spins faster, that also acts as a protection against flipping of the rotation axis. So, perhaps if the Moon had not come off the Earth, our world would still be spinning fast enough to stabilize itself against flipping. Thus there may be many other habitable planets without a large moon, but the inhabitants will have even fewer hours in their day than we do.

The Moon, of course, is now perfectly placed to exactly cover the solar disk during eclipses. This perfect fit has allowed, for example, a test of General Relativity, the uncovering of the element helium, and the discovery of the solar corona. And clearly, the Moon has been a great stimulus and practice ground for our first efforts at space travel. However, moving away from Earth at one inch per year, in about 1.6 billion years, the Moon will no longer be able to stabilize our planet's spin. We'll have to be ready for a climatologically wild ride by then unless we figure out what to do. Eventually, the Earth will have the same rotation period as the Moon's orbit (i.e., the day will equal the month), and then the Moon may be expected to fall back toward the Earth, forming a ring perhaps not dissimilar to those around Saturn. It will, no doubt, be a great show.

Andromeda: Coming to a Galaxy Near You

We could talk about many more interesting phenomena, but perhaps the most spectacular will be the merging of the Andromeda spiral galaxy with the Milky Way in about six billion years. Although no stars will likely touch (the spacing between stars is huge), this interaction will most certainly gravitationally affect every star in both galaxies. The Milky Way, in its 12-billion year history, has swallowed up many smaller galaxies, but such a merger will be a unique experience. Every star is thought to have a cloud (called the

"Oort Cloud" in our solar system) that consists of about a trillion comets. As the two galaxies merge, these comet clouds will get scrambled, causing increased impacts on each star's inner planets. Billions of stellar systems being tidally flung around may also cause instability in the orbits of the planets around them.

Nicolai Kardashev has suggested that there could be three classes of civilizations: Type I – controlling the resources of its planet; Type II – controlling the resources of its star; and Type III – controlling the resources of its galaxy. At the moment, we are estimated to be about type 0.1 or so. A Type II builds such things as Dyson spheres (structures encompassing the star so as to capture all of its energy). Clearly, a Type III civilization would be needed to deal with the merging of Andromeda with the Milky Way.

So, there we have it – some items on the agenda for the next few billion years. We have some time for planning, but it's important that we stay alert. And who knows, we might make it to Type III before Andromeda gets here. If not, perhaps some other species in the galaxy may have gotten it together enough to help us out.

Talking with Moon Men

NASA Hubble James Webb

Although I've been to talks by Alan Shepard – the first American in space, and Apollo 14 Commander – and Buzz Aldrin, Apollo 11 lunar module pilot, and the second man to walk on the Moon – I have actually only had the opportunity to chat at some length with two "Moon Men."

My first experience was over two decades ago. I was a few years old when Sputnik first flew, and I started a scrapbook. That scrapbook contained newspaper clippings of every spacecraft – manned and unmanned – ever flown, from Sputnik to the NASA Gemini two-man-per-capsule missions. I'm not sure why, but I had a favorite astronaut at that time, and it was James Lovell. I followed every flight he had, and gave him a special place in my scrapbook.

When I was finishing a Master's program in Astronomy in San Diego, I heard that James Lovell, of Apollo 8 and Apollo 13 fame, was going to speak, and I showed up early. He walked in the door a half-hour before his speech and no one seemed to notice. I walked over, introduced myself, and we had a chat for about half an hour –just the two of us.

We talked about two things: Apollo 13, and, for a much shorter time, the fact that I was interested in the astronaut program. I was encouraged to hear that a Master's degree in astronomy was "pretty good." But mostly it was Apollo 13. I remember telling him it would make a great movie. As we talked, it began to really dawn on me – this man had been to the Moon! Not once, but twice! Although he did not walk on the surface, it was his command ship that the world had prayed about when Apollo 13 was in trouble.

I don't think the world has ever been as united, before or since; not even when Apollo 11 first landed on the Moon. We had all been such a global community when virtually *everyone* on the planet had been united in support for the Apollo 13 crew to come back alive. It's difficult to describe unless you experienced it. We knew that all religions were praying, meditating, chanting...whatever they did – for these men. From the Pope, to presidents, to royalty, to the folks on the street – everyone that had any access to media, from newspapers, to radio, to television, to gossip, knew about Apollo 13, and they were our – all of our – people up there, and we had to get them back home safely.

Gene Kranz called it "NASA's finest moment" when they got back alive, and truly, in retrospect, it was the finest moment of many fine moments. Well, James Lovell did write the book, "Lost Moon," and it was made into the Ron Howard movie, *Apollo 13*. But the difference between the movie and real life was that everything seemed to happen at once, and the certainty of their safe return was by no means assured. The arriving support crews had left their cars, with the keys still in them, parked all over the roads around Johnson Space Flight Center, and ran into Mission Control, for in some instances they really did have only 15 minutes or so to save the crew.

So here I was, I realized, talking to a man who had actually been to the Moon. We had made it! As I shook Mr. Lovell's hand and he went up to speak, I felt I had gotten the inside story – I'd really gotten it – how important it had been for us to have gone to the Moon.

It wasn't just the spin-offs, to use NASA-ese. Sure, Silicon Valley and the computer revolution would certainly not have been possible without the race to the Moon. And I often think that historians do not give enough credit to John F. Kennedy and Nikita Khrushchev for diverting their military industrial complex's attention to a race to the Moon. There is no telling, especially after the Cuban missile crisis, where the world might have been if the decision had not been made to divert at least some of that competition to something more productive than war.

Khrushchev's son maintains that his father had had an offer from President Kennedy to collaborate in going to the Moon. But within two weeks of that offer, which Khrushchev was apparently ready to accept, Kennedy was assassinated. The Cold War might have been dissipated decades earlier.

December, 1972 is the last time we set foot on the Moon. And one of the last foot prints belongs to the next "Moon Man" I got to spend some time talking with – Dr. Harrison "Jack" Schmitt, Apollo 17 lunar module pilot and the only career scientist (geologist) to set foot on the Moon.

We met first when I heard he was speaking at a meeting of Astrobiology interns at NASA Ames Research Center. I had only heard about it the day before and attended the lecture. Afterwards, I found myself being a groupie, and standing in line to shake his hand. As I stood in line a few from the front, I again had that comprehensive feeling, "My gosh! This guy has walked on the Moon, for Pete's sake!" I smiled and shook his hand and turned to go. I turned back around and just added, "You know, I did my Master's Thesis on the Origin of the Moon."

He smiled a big smile and asked me if I believed the giant impactor theory – that the Earth had been struck by a giant planetesimal the size of Mars, and this is what knocked the Moon off the Earth. I knew almost everyone presently adhered to this theory, but I had not been convinced yet myself, and decided to just tell the truth. I said something like, "Well, actually, I know a lot of good work has gone into this theory, and it has many interesting features... and ah... well, ah... no. I may not be qualified, but I actually don't believe it."

A huge smile. "Neither do I!" he said. A kindred spirit! And a kindred spirit that has walked on the Moon! And a kindred spirit that has studied lunar geology all his life! Well, we compared notes for the next half-hour. Our discussion went on interrupted only by other people filing past to shake his hand. Really great. And we'd have a couple other opportunities to compare notes at scientific meetings as well. When it came to the origin of the Moon, well one could just not find better support on ones side than the only scientist-geologist Moon Man.

And so it has been more than half a century since humans – "we" – walked the lunar soil. It has been difficult for my generation to be mid-teenagers and see folks walking on the Moon, and then just stand by while humans do not leave Earth orbit for generations. It has been difficult to stand by and watch the expertise dissipate, the adventure of exploration be put on the shelf. To just remember the old days. Perhaps we just lost our nerve.

But today, there might be a new Moon race looming. The Chinese already have well-qualified astronauts in Earth orbit now, and have indicated that they'd like to send folks to the Moon within the next few years. As of this writing, only four of the twelve men that walked on the Moon are still with us. All the Apollo astronauts are at least in their late-80s to mid-90s. So we have no available astronauts with the experience of landing on the Moon. We also scrapped the Saturn V rocket, which never failed to get us there, but which we no longer can make, as it was completely de-tooled when we put our eggs in the space shuttle basket, so to speak.

We may have to start from scratch with a lot of this if we are to go back to the Moon again. It's an interesting historical switch. Cheng Ho, under the emperor Yung Lo, sailed several times as far as East Africa, and perhaps into the Atlantic Ocean, with hundreds of ships and thousands of men from about 1405 to 1433. But today, the Chinese language is not widely spoken outside of China. With a new emperor, this actual beginning of the age of exploration ceased, and we generally do not include it in the history of the Age of Exploration.

For a while there, it looked as if the language of space would be English, and then, with the continuously inhabited Mir space station, that it would be Russian. But perhaps the language of the Moon will turn out to be Chinese, in an interesting switch of history. Perhaps the age of continued manned Solar System exploration will begin with the first Chinese expeditions to the Moon, with the earlier US landings an interesting side note, like Cheng Ho's trips to Africa.

But no. Going to the Moon for the first time was not so much like pioneering the wild west (for the European Americans) or the wild east (for the European Russians). It might not be a coincidence that the two space-faring nations had just explored a whole continent each the century before. But going to the Moon was perhaps more like first crawling out of the ocean and trying to live on land. It was something really different for our species; no species from Earth had ever done such a thing before – left our planet completely for another world.

And we, perhaps, in the long term, have no choice about it. I think it is in our nature to overcome the limits of space and time. How great to have such a huge satellite-Moon on which to practice exploration, leading to the planets and the stars. As the 19th century Russian scientist-engineer, Konstantin Tsiolkovsky, is often quoted, "The Earth is the cradle of the mind, but we cannot live forever in a cradle."

Apollo 8 first snapped a picture of the Earth "rising" above the Moon in the late 1960s. It was actually the spacecraft that was rising; the Earth does not rise from any given spot on the Moon, because the Moon is tidally locked in rotation, although the Earth does go through phases opposite those of the Moon – a full Earth happens during a new Moon, and vice versa. But that picture showed us that the Earth is a precious cradle indeed, and the modern ecology movement was launched (coincidentally?) during that same time. I wonder what other lessons await us from the perspective of space? Perhaps it's time we ventured out again and found out. It will be fun to see what the next few years have in store.

The Role of Sponges in the Galaxy

NASA Hubble James Webb

There are many questions of key interest to SETI, for example: Why didn't the dinosaurs go to the Moon? They had 200 million years, and many species had hands with an opposable digit, big brains, and were also bipedal.

Another entirely different SETI question could be: What do Medieval Arabic texts have to say about the origin of optics? Light was thought to come from the eye at the beginning of the Middle Ages, but within a few centuries, advanced optical studies emerged from Arab countries with refracting lenses and prisms, and light was understood clearly to go into, not come out of, the eye. Such an emergent process would be essential to understand if one wants to generalize the development of telescopes by intelligent civilizations.

Another interesting SETI-related question: When considering the communication systems of more advanced technical civilizations, what can we expect their communication system to be like? Since something like 98% of the stars in the solar neighborhood are apparently older than the Sun (and for other reasons), we can likely expect that an extraterrestrial technology will be significantly more advanced than ours. As an example, one can hardly ignore recent developments in quantum teleportation as a possible neat trick for instantaneously getting large amounts of information across the galaxy. It may or may not be possible to then read that information faster than light, but that's another story.

These are all interesting SETI questions and might make good articles themselves. But for this essay, let's ask ourselves this SETI-related question: Would the development of our technical civilization have been possible without sponges? I shall not so much answer this question in our essay as explain why we I asked it.

Given that a tool-wielding species has to start somewhere, humans apparently started chipping rocks to form cutting surfaces at least 4 million years ago. We see wild chimpanzees doing this today; they also train their offspring to do this as well, so it is a learned, or cultural, behavior. But as it turns out, banging special kinds of rocks together can produce an additional feature of technology, one that would come to distinguish us as a species about 2 million years ago. It all started when we banged together iron pyrite with ocean sponges – actually, to state it more properly, we started to bang together iron pyrite containing a metamorphosed mixture of chalk with the internal skeletons of ocean sponges. This metamorphic chalk and sponge skeletal material is commonly known as flint. And, as we know, when struck with iron, it makes a spark of fire. And we are the only species that uses fire. (I'm not counting an entertaining raven that actually does a nightclub trick in Las Vegas that includes striking a match to light a cigar.)

So, although flint is an inorganic mineral, almost all the silicate in it is derived from the dissolved skeletons of sponges. This strike-sparking is considered the earliest form of fire-making. Later, steel would be substituted for the iron pyrite, but flint was used for millennia before the invention of matches in the 19th century. And it would be difficult to argue for a more important invention to the survival of the human species than the making of fire – especially during the frequent ice ages that accompanied early northward migrations out of Africa. It could also be argued that fire-making was the first big step of our species toward a technical civilization.

It turns out, then, that the process of sponge skeletons dissolving in pre-lithified chalk ooze – dehydrated and hardened into microscopic quartz crystals forming flint – was essential (along with trees, but that's another story) to the survival and eventual technical success of our species on this planet. It could be argued that without sponges, humans would not be listening to or transmitting messages over interstellar distances today.

So, looking for extraterrestrial technology? Then it might be a good idea to know where in the galaxy the sponges are. The humble sponge, or something like it, may be essential for any species to first come up with the making of fire. And on our planet, if you don't have a dishwasher, they still come in pretty handy too.

Ancient African Skies

Photo by; Tusker Trail, Eddie Frank

"Bwana! That must be it!" I pointed back over my right shoulder at some man-sized stone pillars off on a rise from the dirt road we were driving on. The road would be described by Kenyans as "corrugated," meaning that we had to get out in some places to look for it on our way to the Turkana region – not too far from where the most ancient hominid fossils had been found by Richard Leakey, Director of the National Museums. We had heard of an ancient astronomy site in this region made up of basalt pillars that were magnetic, so it needed to be re-measured using something other than compasses. Eddie Frank of the Tusker Trail Company, some students from various places in the U.S. and I had come to do just that. We had camped with hippos at Lake Naivasha, swam in Lake Turkana, which is known to have over three million crocodiles in it, watched as several million pink flamingos took to the air around Lake Nakuru, and seen the migration of a couple million wildebeests across the Maasai Mara plains, all on our roundabout safari to this ancient astronomy site.

The pillars were known as "Namoratunga," or "stone people" in the Turkana language. They had been said to have been built for an astronomical purpose – to reckon the Borana calendar – and were reputed to be a couple of thousand years old. They had petroglyphs on them, some matching the ancient property symbols of the Kush, a people of the Sudan who conquered Egypt in the 8th century B.C., and whose language has yet to be deciphered. From maps of ancient Kush, I knew that the front of almost all their pyramids appeared to face the star Sirius, which has not changed its position in the sky very much for the last several thousand years.

We had a description of the Borana calendar, but this did not make astronomical sense. For example, the new year of the Borana calendar was said to occur when the star Beta Trianguli was "in conjunction" with the new Moon (give or take a couple of days). A new Moon is, of course, close to the Sun (i.e., a thin crescent), and consequently found in the twilight sky. However, the star Beta Trianguli is a 3rd magnitude star and could not been seen in twilight. Thus, we could not even get the Borana New Year started according to the description we had from anthropologists.

We made camp under some acacia trees and the next day tramped over to the ancient site. It was obvious that the stone pillar had been polished by the wind over a long time. We used navigational sextants to measure the angles between each pillar from the others, and a measuring strip to map the distances between them. The day was blessedly cloudy – the first such day in the Turkana desert we had seen. We sketched the petroglyphs on each of the pillars as well. And Matunga, our cook, brought us all lunch at the site with some very good Kenyan tea. Sitting among the pillars, it occurred to me that perhaps the translation of the word "conjunction" might have been incorrect. What had been taken to mean "rising with" on the same place on the horizon, might instead mean "rising single-file after." Hmm, that could account for having pillars also line up with the stars and the Moon.

There are seven special Borana calendar stars that define six places in the sky: beta Triangulum; Pleiades; Aldebaran; Bellatrix; central Orion and the star Saiph taken together; and Sirius. Well, if the single file idea was correct, then at the beginning of the Borana year, the star beta Triangulum would rise – mark the place with a pillar, for example – and then at dawn, the new Moon would rise in the same location on the horizon. We were at 3.4 degrees north latitude, so the stars there rise almost vertically; 3.4 degrees off the vertical to be exact. However, this vertical, single-file "conjunction" also would not work because here, very close to the equator, the star beta Triangulum was rising about 35 degrees north of east, but the farthest north the Moon ever gets in its 18.6-year cycle (called "regression of the lunar nodes") was 28.5 degrees north of east. So, again, the calendar could not even get started. What kind of a calendar was this, I thought? When we had met with the scientists at the National Museums of Kenya they also did not know how the calendar worked. Yet the Borana are practical people; they would not have made up an apparently ancient calendar that didn't work.

Ah, ancient! I knew that while the star Sirius had not moved much, many of the stars had altered their apparent position on the sky. This "precession" is due to the wobble of the Earth's rotational axis over a complete circle on the sky every 26,000 years. During the time the first pyramids in Egypt were built, the north star was not Polaris in the Little Dipper, but Thuban, in the constellation of Draco the Dragon. This is where several openings in the great pyramids were aligned. The Namoratunga site was said to possibly be as old as 300 B.C., based on some carbon dating of bones. We precessed the Borana stars back to 300 B.C., and some "moved" quite a bit. But where was beta Triangulum in 300 B.C.? Before I answer that question, let's take a quick look at how the Borana calendar is actually supposed to work.

There are no weeks in the Borana calendar; our seven-day week came from the six "planets" (Venus, Mercury, Mars, Jupiter and Saturn, plus the Moon) and the Sun, all thought in ancient times to circle the Earth. "Saturday" came from Saturn, "Sunday" was named for the Sun, and Monday's name came from the Moon. In English calendars, we use the names of Norse gods for the other days – Tuesday is named for Tyr, Wednesday for Odin, Thursday is derived from Thor, and Friday comes from Freyja.

The Borana have 27 day names and 12 lunar month names. The first day of the New Year starts on the day named "Bitto Tessa" in the month "Bita Kara." This is when Triangulum is in "conjunction" with the new Moon. One then simply counts the day names through the month based on that first astronomical observation – conjunction of the new Moon at the beta Triangulum position. The description continues; one will know that the next month begins when the new Moon is "in conjunction" with the next star or star system, in this case, Pleiades (a blue star cluster). This occurs 29.5 days after the start of the first month. That means one runs out of day names a couple of days early. This is OK, and the names of the days that started the month are also the names of the days that finish the month. This is the same for all the months, adjusted for the observations, of course, allowing a variation of a day or two here and there based on the astronomical observation.

The third month starts when one spots the new Moon rising "in conjunction with" the star Aldebaran, and so on down the list of six Borana star positions for the first six months. Why does this work? It is because the "sidereal month" (the time it takes for the Moon to move from a certain star position back to that position again – 27.3 days) is not the same as the "synodic month" or the time it takes the Moon to go from a particular phase (full Moon to full Moon phase, for example) back to that phase again (29.5 days). The synodic month is longer because the Earth has orbited the Sun just a bit also in a month's time and the Moon has had to (sort of) "catch up." That is, when the Moon arrives at the same place again (star position), the Earth has moved a bit farther and so to align with on the exact opposite side of the Earth from the Sun (same lunar phase again) the Moon has to travel a bit farther. (Try this if you like with three people – one the Moon, one the Earth, and one the Sun, with the Sun shining a light onto the others they move around the Sun and Earth, respectively; see what happens.)

So, back to the 300 B.C. positions of the stars – would the calendar work if we put the stars back to the same time as the pillars were expected to have been put there? Indeed, the second new Moon rose at the exact place where Pleiades used to rise in 300 B.C.! An exciting moment! The next new Moon? It rose at the 300 B.C. position of Aldebaran! The next? Bellatrix! The next? In between central Orion and the star Saiph! And finally the 6th new Moon of the year rose where the star Sirius rose (and still rises) on the horizon! It worked in 300 B.C.!

Well, it looked as if we had deciphered an ancient calendar! The next six months of the year are defined by a switch of the moving parts. Instead of the horizon location changing, the next six months are defined by the various phases of the Moon rising at the Triangulum-only position (and therefore one can only check the calendar in the middle of the month). But this all worked too. One goes through the full Moon, three-quarter (gibbous) waning Moon, quarter waning Moon, large crescent waning, medium crescent waning, and finally small crescent waning – all rising at the Triangulum position until one is back to the new Moon at the Triangulum position again and Happy New Year! The exception is that every three years another month is added because the lunar year is 11 days short of the solar year; this is a kind of "leap month" the Borana folks used.

So what about the 19 stone pillars that make up Namoratunga? Are they indeed used for the ancient Borana calendar? We found they made 25 alignments with the seven positions on the sky of the ancient Borana stars or star systems. After some calculations, we found that this many alignments could only have occurred randomly about 43 times out of 10,000 random star positions, counting the alignments with the pillars generated randomly by computer, and doing this test of randomly made-up star position 10,000 times for good measure. In other words, we found also that the pillars at Namoratunga, to have made as many as 25 or more alignments with the seven specific ancient positions of the Borana stars, would only occur randomly 0.43% of the time. From this mathematical experiment, we could be about 99.57% sure that the pillars at Namoratunga had been built to do just that – line up with the Borana calendar star positions for 300 B.C.! And the petroglyphs on the pillars? The Turkana said they were ancient family names. This was interesting too, because a lot of them match the symbols on the pyramids of the ancient Kush people; symbols said to be royal property symbols. Perhaps the Kush also got as far south as they did north in ancient times. And perhaps petroglyphs might be a way (besides linguistics and genetics) to trace the migrations of ancient peoples.

As we began the long drive back to Nairobi, I felt a quiet excitement. We had discovered a way of keeping time and one of the ways of thinking of some of the ancient African astronomers who had lived in the Turkana desert over 2,000 years before. They looked at the sky and understood how it worked, and with the stars and Moon as intermediaries, we had shared this special order with them across the millennia. We had read the same book of the sky together, and the practice of timekeeping that so often seems to separate us had, instead, this time brought us closer together.

You can read a more detailed description of the Borana Calendar and Namoratunga in "The Encyclopedia of the History of Science, Technology and Medicine of Non-Western Cultures," edited by Helaine Selin, in the chapter "Astronomy in Africa," by L.R. Doyle and E.W. Frank.

SETI Evolution: Searching for Aliens Using Whale Songs and Radios

The Search for Extraterrestrial Intelligence (SETI) can be said to have begun in 1896, with the suggestion by Nikola Tesla, designer of our modern, alternating current electrical system, that radio transmissions could be used to contact an extraterrestrial intelligent being. In 1899, Tesla actually did detect signals incompatible with, for example, terrestrial electric storms, but some have suggested, after looking at Tesla's data, that he might have been picking up "storms" on Jupiter (the Jovian plasma torus emits strong radio flux, making Jupiter a kind of miniature pulsar).

Later, in August 1924, when Mars was in inferior conjunction (closer to Earth than it had been in over a century), the U.S. Naval Observatory imposed radio quiet for five minutes out of every hour at so that a dirigible equipped with a radio receiver could listen for any Martian signals.

But modern SETI really began in 1959, when Philip Morrison and Giuseppe Cocconi wrote a paper for the journal *Nature*, pointing out that extraterrestrial intelligence might be detected using radio antennas. Independently, in 1960, Frank Drake used a radio telescope to conduct the first SETI project by looking at two stars, Tau Ceti and Epsilon Eridani, for signals being sent from extraterrestrial life using the electromagnetic frequency of 1.420 gigahertz with a 400 kilohertz bandpass, meaning about 400,000 different SETI channels fit in a region of the spectrum this wide. That frequency is in what is known as the "water hole," a range of frequencies that water vapor does not absorb very well, and so a water-based planet might transmit signals into space at such frequencies.

Drake developed a way of organizing the search for extraterrestrials using a tool now known as the "Drake Equation." The equation reads $N = R_* \, f_p \, n_e \, f_l \, f_i \, f_c \, L$, where N is the number of interstellar communicating civilizations, and the other terms have the following meanings:

The term R_* – the number of good, or habitable, stars – usually means solar-type stars, but research over the past couple of decades has shown that smaller, red dwarf stars can also be good locations for habitable planets. However, for planets from those stars to have liquid water, they must be at the right distance from their stars and tidally locked in rotation (one side of the planet always faces the star). Red dwarfs being 75 percent of the stellar population, those results have significantly increased SETI target stars.

The term f_p is the fraction of habitable stars that actually have planets. Since its launch in March 2009, the NASA Kepler telescope has actually determined the frequency-of-occurrence of planets of various sizes. It is a remarkable achievement and a very important one for the SETI efforts. At the SETI Institute, we have targeted and "listened to" all the detected planets that lie within a habitable zone. The primary goal of Kepler is to detect Earth-sized planets lying within the habitable zones of their stars. This is an ongoing process at the moment, and depends on the repair of Kepler's reaction-wheel gyro so it can point accurately.

Finding an Earth-like planet in the habitable zone would give us the Drake Equation term n_e, the number of planets that are Earth-like (in terms of size and location) within a given stellar system.

The next term would be f_l, the fraction of potentially habitable worlds where biology actually gets started. For this, detection would require a new generation of orbiting telescopes able to detect, for example, oxygen in the atmospheres of habitable planets, a sign of photosynthetic organisms. The first extraterrestrials to be detected might be forests; they have been around on the Earth for over 400 million years, while radio has only been around for a bit over 100 years.

The next term is f_i, the fraction of biological organisms that develop intelligence. This is the most difficult term of the Drake Equation to define. It raises more fundamental questions, like, "What is intelligence?" There are many kinds of intelligence, but for the purposes of SETI, it seems that this would mean communication intelligence, because this is what humanity would likely encounter – i.e., a communication.

If one wants to find exobiology, one might start by studying the extremes of biology on Earth (e.g., the NASA astrobiology program). So, investigators go to the dry valleys of Antarctica or the deserts of the Mojave in California to see how tough biology is.

Similarly, it seems to me, if one wants to detect non-human communications from space, one should start studying the multitudinous non-human communication systems here on Earth. All animals (and plants, too, for that matter) communicate. But how do scientists analyze and determine the complexity of these communication systems, assuming, for now, that communication intelligence can be measured by message complexity?

Can researchers determine how complex these communications are? A tool for measuring message complexity lies in a mathematical field developed initially to measure how much information is transmitted through telephone lines, and it is called Information Theory. This discipline was developed by Claude Shannon of Bell Laboratories in 1949, and is used extensively today – for example, if you have zipped or unzipped a mailed file, you (or your computer, anyway) used information theory.

With colleagues Brenda McCowan and Sean Hanser of the University of California, Davis, we decided to apply information theory to bottlenose dolphin communications to see how much information they were transmitting to each other using their whistle communication system. That information amount depends on the distribution of the signal frequencies-of-occurrence and is called the Information Entropy.

One early aspect of that kind of analysis is known as Zipf's Law, named after the linguist who plotted the logarithmic-scale frequency-of-occurrence (from most frequent to least frequent) of English letters in novels. He obtained a more-or-less straight 45-degree line, which is a slope of -1. In other words, the most-frequent letter occurred 10 times more frequently than the second most-used letter; the third most-used letter, one-tenth as often as the second, and so on, from the letter "e" and then "t" to the least-used letter, "q." He also plotted this for Chinese characters, English words, Russian phonemes and so on, and always got a line with a -1 slope drawn through the frequency points.

It turns out that Zipf's Law appears to describe a distribution of signals that is necessary to make a language. However, it is not a sufficient characteristic, as other processes can produce this -1 slope, as well. We did a Zipf plot for bottlenose dolphins and got a -1 slope. This meant that their whistle communication system could contain complicated relationship rules within it (what linguists call "syntax" in human communication systems). It turns out that the Zipf slope for baby babbling is a lot less steep than Zipf's Law, but when baby dolphins were born at the facility where we had recorded the adult dolphins, we recorded them and found that the distribution of their whistle frequencies was the same as baby humans. We see both growing into their -1 language use. This means that baby dolphins "babble" their whistles when they are little and learn their language as they mature.

We also applied these techniques to humpback whales, in part because they are also a socially complex species, and in part because they rely – as dolphins do – on vocal communication more than gestural or facial expressions. This species also had a global communication system millions of years before humans did.

These animals also resembled humans in the way they dealt with noise. When one is talking on a phone line with static (or "noise," as engineers call it), the speaker has to slow down the rate of words in order to ensure that the listener gets all the words. We also found this to be the case for humpback whales in the presence of boat noise – they slowed down the rate at which they transmitted signals to each other. They were communicating to each other while making bubble nets to trap herring. We could calculate how much the whales would have to slow down to compensate for the boat noise, but it turned out that they only compensated by about 60 percent – in this case, they did this by increasing the total time they used to transmit signals to each other (i.e., they slowed their speech).

So what was going on here? An analogy can help. When your copy machine is low on toner and you don't want to make another copy, you can fill in missing letters and words using the rules of spelling and grammar. This works as long as you are not missing too many words in a row, because if that happens you cannot use rules and context to recover missing words. It turns out that, for words, humans have a non-zero probability for word recovery with up to nine words missing in a row – known as the "ninth-order entropy."

We realized that the humpback whales must also have a kind of "syntax," because they got the gist of the message without having to hear it all the way through. We don't yet have enough data to determine if the information entropy (complexity or "rule" structure) is as high as ninth-order entropy, as is the case with humans, but we do know that whale signals have "condition probability" between them. This is a technical way of saying that these animals have structural complexity in their communication system.

I've also applied information theory to a one-way communication between cotton plants and wasps, the cotton plants telling the wasps on which plant to land (the plants that have the worms the wasps are targeting). This is not an inter-world communication, but perhaps as close as scientists can currently get to an inter-kingdom (plant to animal) communication system.

I think it would be interesting to apply information theory to honey bees next, as their Waggle Dance communication system involves using the Sun for navigation. Bees have the three important ingredients necessary to be identified by a SETI analysis – a communication system, tool use (building the hexagonal hive structures) and astronomy (they use the Sun and sometimes the full Moon to find honey sources).

In addition to radio searches, optical SETI is now becoming more widespread. Radio SETI is looking for narrow-band transmissions (where one can turn the dial of the radio once and be on a new station), which nature apparently cannot produce. Optical SETI, however, relies on the detection of nanosecond pulses of light. Again, technology can produce such signals, but, as far as scientists know, nature cannot (the quickest pulses in nature may be millisecond pulsars).

Would an advanced civilization use nanosecond optical (or infrared) pulses to signal over interstellar distances? It might, especially if it has a nice laser handy in its solar system. In the Earth's solar system, there is actually a naturally occurring microwave laser, or "maser," in the atmosphere of Mars. That planet's atmospheric carbon dioxide regularly "mases" with input from the Sun. The upshot of this is that if one surrounded Mars with mirrors to focus this natural masing action, one could indeed send interstellar signals, essentially for free. Humanity's own civilization needs a few more decades (or maybe a century) before it can utilize this method of communication.

Other SETI folks are searching for excess heat from Dyson spheres. Named after Princeton University physicist Freeman Dyson, who initially thought them up, these would be spheres of satellites or other structures that a civilization would place around a star to capture as much energy from it as possible. The spheres would have a radius about that of the Earth's distance from the Sun. The interior of such a sphere would house a lot more people (or aliens) than little planet Earth could handle, as it would have something like 550 million times the surface area of the Earth.

Another interesting proposal for SETI is based upon the notion that biology on Earth did not actually begin on Earth. Some people have suggested that biology originated on Mars, and that cometary impacts on Mars brought very simple bacterial organisms to Earth. Interestingly, if this were true, people from Earth might really be from Mars!

But a new SETI idea is even farther out than that. The idea is that there is a SETI-type "calling card" in the human genome. In order for this to be isolated, one would have to show that this particular region in the human (or perhaps another species') genome was not just non-random (any process with a rule structure of any kind is non-random), but that this certain region of the genome was incompatible with the processes that shaped or altered the present genome. The idea is that if a region of the human genome could be shown to not be like any other parts of the genome, and – much more difficult – to not be producible by natural selection, for example, then it would have to have been made by a pre-human and very advanced intelligence. I think information theory here would be very useful, as one could perhaps isolate regions of the genome that had unusual structure.

When pondering SETI efforts, one must consider what very advanced civilizations might be doing. One can hardly, therefore, avoid the latest in information transfer – quantum teleportation. So far, research in quantum optics laboratories has shown that information can (it appears) be instantaneously teleported over arbitrary distances. But it should perhaps be said, more precisely, that "quantum information" can be instantaneously teleported arbitrarily far away, because one must then use a kind of "key" to unlock the information. And the key cannot be transmitted faster than the speed of light. Nevertheless, one wonders if the first SETI message is going to come from space or appear in one of our quantum computers.

The search for extraterrestrial intelligence is a fascinating field and encompasses the whole gamut of disciplines, from astrophysics to animal communications, and from paleontology to quantum mechanics. At present, it is symbolized by a radio telescope, but that's just the tip of the iceberg.

In a book on life in the solar system called "Other Worlds Than Ours," published in 1870, a British gentleman named Richard Proctor highlights a 17th century French work called the "Plurality of Worlds," and discusses the habitability of other planets. He stated that most of the public's interest in astronomy derives from its interest in finding life in space. I don't think this has changed.

The Legend of Rongo Rongo

Rongo Rongo

Image Credit: Dr. Fred Sharp & Sean Hanser

We had just landed at the Juneau, Alaska airport – our last sign of what is called "civilization" for a while. Juneau is the only state capital that one cannot get to by road; one either flies or boats in. Along with colleagues Dr. Brenda McCowan of UC Davis and Jeremy Crandell of PlanetQuest, we walked toward one end of the airport and then outside to encounter what might best be described as an "Indiana Jones" type airplane. It was a biplane on pontoons, below which were retractable wheels. We were flying to the small Native American Tlingit village called "Angoon," to meet up with colleagues Sean Hanser of UC Davis and the Alaska Whale Foundation on their research vessel, the "Evolution." I actually got to sit in the co-pilot's seat, as the pilot needed someone with a strong arm to retract the wheels by hand soon after we took off.

We had come to study the humpback whales, specifically to record their social and working vocalizations, possibly the most complex in the animal kingdom. We have recordings of humpbacks that sound just like wolves howling, monkeys barking, dolphins whistling, cows mooing, lions roaring, elephants trumpeting and even humans talking, along with numerous additional bubbling, gurgling, screeching, groaning and burping sounds, unique to their own repertoire.

One of our projects was to record humpback whales in the presence of boat noise, and compare it with their vocalizations in the absence of boat noise in order to determine if boat traffic could be interfering with their feeding and social behavior. For this, we were using hydrophones and information theory analysis software we had developed based on Shannon Information Theory – the mathematics of quantifying how much information is being sent across a given phone line. Usually used to quantify how much information is being sent through computer lines these days, we instead just assumed that Chatham Strait itself is a communication channel and that the humpbacks are sending information to each other at rates that might vary with and without background noise.

I, of course, was interested in the generalized communication complexity of the whole animal kingdom for the purpose of formulating general rules for application to any possible SETI (search for extraterrestrial intelligence) signals that may someday be received. However, being in the presence of working humpbacks certainly helped give me a sobering thought about any possible other species on another planet. When a couple of dozen humpback whales surface, one is inexorably led to consider how really alien their world is compared to our own daily lives.

Unlike the humpbacks in Fredrick Sound, for example, which generally eat krill (which they just guide into their open mouths), the humpbacks in Chatham Strait go after herring, which can swim faster than the humpbacks can. But they have a trick that must certainly qualify for bona fide tool-making. They coordinate with sounds, and some individuals start to make a net of bubbles while another (or others) start to herd the herring in the direction of this cylindrical net of bubbles. Timing is crucial, as the net, being made of bubbles, is rising toward the surface all the time. The herring are herded into the net, and the humpbacks come up underneath with their mouths open. Each humpback mouth can hold a bit more than about a moving van's worth of water and fish. It is an unforgettable experience to see them break the water with huge gulps and trumpeting exhales.

In addition, from mitochondrial DNA analysis done by Dr. Fred Sharp and Pieter Folkens of the Alaska Whale Foundation, these fishing groups of up to two dozen individuals are not directly related to each other. This is of interest, as it may represent the only other long-term, non-familial associations beside humans, e.g., humans associated by given professions. "Tryouts" for the fishing team have been observed, and one can hear when the bubble net has been successful or not. The vocalizations proceed for a while, then one hears a rise in tone, and then one looks for a huge amount of bubbles, followed by fish flying out of the water, followed by huge open mouths breaking the water. Once a fishing party is formed, all two dozen of the 50-ton animals always come up the same way, time after time. They all know their place in the fishing scheme.

When one hears these vocalizations, but then right in the middle of the sequence they stop, it means a bubble net has not been built correctly, or a herder has not done their job correctly, and the herring have gotten away. This is followed first by a slow rising of the humpbacks, followed by a whole bunch of vocalizations. One can almost imagine that they are arguing about who messed up and why. Following this behavior, it has also been observed that one or two big humpbacks fill their mouths with water (to be bigger), and push an individual or two away from the pack, excluding them from participating in the bubble net fishing group. Then they head back down again and the whole process starts over.

The Alaska Whale Foundation has a catalogue of over 500 individual humpback whales, identified by tail fluke (but after a season one can tell them also by their vocalizations). Each has a name – Melancholy, Beethoven, Taylor, etc. While on the boat, we were talking about information theory a lot, and we had recently applied it to the analysis of the undeciphered and only indigenous written language of Polynesia, known as "Rongorongo." Using information theory, we had discovered something about the characters in Rongorongo that might be helpful in its interpretation. So, when a brand new humpback whale showed up this last season, the Alaska Whale Foundation folks decided to name the newcomer, "Rongo Rongo." He has quite a unique fluke design – almost two "W"s can be seen (with a bit of imagination) on each fin (see picture).

So, I guess it is a bit early to tell yet if Rongo Rongo will become a legend or not. But he was the first new humpback whale to show up while we were applying information theory to humpback vocalizations. And who knows? He may turn out to make a major contribution to the analysis and possible decipherment of an extraterrestrial signal someday. For now, he is honing his fishing skills – trying out for the Bubblenetters Guild, no doubt – and joining his colleagues in making the most diverse sounds in the animal kingdom. I look forward to seeing and hearing him next season. And it may be that we'll find out that humpbacks and humans are not so alien to each other after all.

A Scientific Referee's Guild

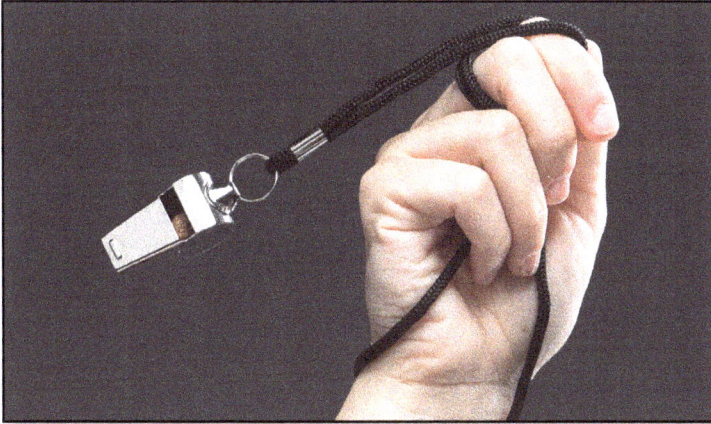

Photo by; @freepik

In this article, I'd like to talk a bit about my own experiences, not in scientific research but in science funding. I don't want to appear to complain; I have no complaints, as I have been able to pursue many different interesting fields of research, and the fiscal support, while dicey at times, was always there at some level. But integrity compels me to write about the process the scientific community calls "refereeing," in which other scientists judge one's work as worthy of publication or worthy of funding, for example, which is the topic I'll discuss here. I have not had many problems getting papers published in the scientific literature.

Since this is not a complaint article, I won't name the one or two funding agencies to which I have submitted proposals for funding, and have experienced not just disappointing results, but disgraceful refereeing. But I can give bona fide examples which I feel on these occasions would rightly call for a Referee's Guild; a body of experts in no way connected with the funding of the projects being refereed, and (ideally) high above reproach. And, unlike today's referees, they would be paid by the publication or program using their services.

From my experience, the need for such a guild has stemmed from mistakes in how referees are picked. For example, in one agency program, the referees were almost exclusively picked from the pool of those who had also submitted proposals. The director of this program argued that this did not produce poor refereeing. I thought it did, and I never got funding from this program even after about two dozen proposals. However, I must say that the local programs of this agency almost always funded me with what is known as "seed" money. But the quality of the refereeing from this program you can judge for yourself. I got the following responses in the early 1990s from proposing to search for extrasolar planets using the transit method – again, not one of these proposals received funding for the following reasons.

"Astrometric interferometry is the only way to detect planets." This one-sentence rejection was given before the first extrasolar planets were discovered using pulsar timing, and then, radial velocity measurements.

Another reason for rejection: "This project can only detect brown dwarfs, not Jupiters." A whole committee let this get through. We were proposing the transit method, which measures the drop in light of a star as a planet moves in front of it. Since brown dwarfs are *smaller* in size than Jupiter-sized planets, and transits measure the drop in light produced by the area of the transiting object, it was absurd to say that a smaller object could be measured and not a larger one.

Another: "This work is already being done by the TEP (Transit of Extrasolar Planets) Network." This was true, but I had to point out to the agency that I was the Director of the TEP Network and so was well aware of the work being done, and could we please obtain some funding to do it? We were all volunteers at the time; I was making my living teaching at a small college and observing in the summer. The reply was basically that it looked like the work was already being done without funding, so why should they fund it? A wry sense of humor is needed for this kind of response.

Anyway, you get the idea. There are many other examples, some less ridiculous, some more so. OK, if you insist, I will share my favorite. I got a referee's report once with the first reason for rejection as, "It's probably going to rain at the observatories." Very wry sense of humor! There were three observatories involved in the proposal, so my colleague immediately calculated (assuming that "probably" meant at least 51%) that it would have to rain about 80% of the time at each observatory, and rightly concluded that no one would build an observatory in such a place. But so much for logic.

I decided that I just had to track this one down. When I finally found the scientist who had written the original referee's report to the agency, he told me he had recommended the project for funding. He had written, "This is one of the best proposals I have read; the only way it can fail is if it rains at the observatory." So, in this case, it was clearly not the scientist's refereeing that had denied us funding but the grasping for negative reasons for rejection by the program itself. Human nature – endlessly amusing, no?

Hopefully, then, we have gotten to this part of the essay without you thinking that I am holding any grudges against any science programs – that's certainly not the case. While disappointing at the time, I also just had to shake my head and sometimes even laugh out loud. Ah, human nature...bless its heart. It is also confidence-assuring that these proposals, almost universally, are now funded programs – not to me, but to various other people and institutions. So at least we know we were in the right scientific ballpark. My institute has called this effect "being ahead of the funding curve." But I'd like to think that scientific referees could, and would, know good science when they see it. After all, science is supposed to be subject to objective judgment as to its correctness or not.

At the moment, as I mentioned, papers for journals, as well as proposals for funding, are refereed by scientific peers. It is understood that this is something one must do voluntarily, and many times (at least in my experience) one has to put in a significant amount of time to do a good job of it – sometimes even a significant fraction of the time it took to produce the original paper itself, as I have also heard colleagues mention. A good start would be having the names of the proposal's authors withheld, while the names of the referees would have to be known, so they could stand by their reputations.

So, although I don't know how practical such a suggestion may be, I would like to nevertheless suggest here that a Scientific Referee's Guild be established, consisting of professional (i.e., paid) referees with the highest credentials. Perhaps they might only be called upon in disputed cases, but it might be good to know such a body exists to get an impartial scientific judgment in such cases. They could hear cases (like the brown dwarf versus Jupiter-like planet detection mentioned above) and intervene to point out that a scientifically valid reason would have to be given to justify rejection of funding, for example. The whole goal of such a guild would be to insure that intellectual integrity and mutual respect be maintained in the scientific community.

Do I see such a guild being established very soon? Perhaps not. But perhaps just bringing up the subject has helped to focus thought on this issue, encourage those scientists whose funding experience might be going through a similar phase, and strike a note for scientific integrity over funding considerations when push comes to shove. I know that this last point may have caused some to gasp, but I remind those scientists out there that they certainly did not go into science primarily to make money. And to remember why they went into science in the first place is to remember the ideal that they must have had in the beginning – that life is primarily an adventure of discovery, and that science itself must be the final arbiter of truth.

Quantum Astronomy, Part I: The Double Slit Experiment

This is a series of five articles, each with a separate explanation of different quantum phenomena. Each of the five articles is a piece of a mosaic, and so every one is needed to understand the final explanation of the quantum astronomy experiment we propose, possibly using the Allen Array Telescope and the narrow-band radio wave detectors being built by the SETI Institute and the University of California, Berkeley.

With the success of recent movies such as "What the &$@# Do We Know?" and the ongoing – and continuously surprising – revelations of the unexpected nature of underlying reality that have been unfolding in quantum physics for three-quarters of a century now, it may not be particularly surprising that the quantum nature of the universe may actually now be making inroads into what has previously been considered classical observational astronomy. Quantum physics has been applied for decades to cosmology, and the strange "singularity" physics of black holes. It is also applicable to macroscopic effects such as Einstein-Bose condensates (extremely cold conglomerations of material that behave in non-classical ways), as well as neutron stars and even white dwarfs, which are kept from collapse not by nuclear fusion explosions, but by the Pauli Exclusion Principle; a process whereby no two elementary particles can have the same quantum state and therefore, in a sense, not collapse into each other.

Well, congratulations if you have gotten through the first paragraph of this essay! I can't honestly tell you that things will get better, but I can say that to the intrepid reader that things should get even more interesting. The famous quantum physicist Richard Feynman once said, essentially, that anyone who thought he understood quantum physics did not understand it enough to understand that he did not actually understand it! In other words, no classical interpretation of quantum physics is the correct one. Parallel evolving universes (one being created every time a quantum-level choice is made), faster-than-light interconnectedness underlying everything, nothing existing until it is observed – these are a few of the interpretations of quantum reality that are consistent with the experiments and observations.

There are many ways we can go now in examining quantum results. If conscious observation is needed for the creation of an electron (this is one aspect of the Copenhagen Interpretation, the most popular version of quantum physics interpretations), then ideas about the origin of consciousness must be revised. If electrons in the brain create consciousness, but electrons require consciousness to exist, one is apparently caught in circular reasoning at best. But for this essay, we shall not discuss quantum biology. Another path we might go down would be the application of quantum physics to cosmology – either the Inflationary origin of the universe, or the Hawking evaporation of black holes as examples. But our essay is not about this vast field, either. Let's discuss the scaling of the simple double-slit laboratory experiment to cosmic distances, or what can truly be called "quantum astronomy."

The laboratory double-slit experiment contains a lot of the best aspects of the weirdness of quantum physics. It can involve various kinds of elementary particles, but for today's discussion we will be talking solely about light, the particle nature of which is called the "photon." A light shining through a small hole or slit (like in a pinhole camera) creates a spot of light on the screen (or film, or detector). However, light shown through two slits that are close together creates not two spots on the screen, but rather a series of alternating bright and dark lines, with the brightest line in the exact middle of this interference pattern. This shows that light is a wave, since such a pattern results from the interference of the waves coming from slit one (which we shall call "A") with the waves coming from slit two (which we shall call "B"). When peaks of waves from light source A meet peaks from light source B, they add, and the bright lines are produced. Not far to the left and right of this brightness peak, however, peaks from A meet troughs from B, because the crests of the light waves are no longer aligned, and a dark line is produced. This alternates on either side until the visibility of the lines fades out. This pattern is simply called an "interference pattern," and Thomas Young used this experiment to demonstrate the wave nature of light in the early 19th century.

However, in the year 1900, physicist Max Planck showed that certain other effects in physics could only be explained by light being a particle. Many experiments followed to also show that light was indeed also a particle (a "photon"), and Albert Einstein was awarded the Nobel Prize in physics in 1921 for his work showing that the particle nature of light could explain the "photoelectric effect." This was an experiment whereby low energy (red) light, when shining onto a photoelectric material, caused the material to emit low energy (slow moving) electrons, while high energy (blue) light caused the same material to emit high energy (fast moving) electrons. However, lots of red light only ever produced more low energy electrons, never any high-energy electrons. In other words, the energy could not be "saved up," but rather must be absorbed by the electrons in the photoelectric material individually. The conclusion was that light came in packets – little quantities – and behaved thus as a particle as well as a wave.

So, light is both a particle and a wave. OK, kind of unexpected (like Jell-O) but perhaps not totally weird. But the double slit experiment had another trick up its sleeve. One could send one photon (or "quantum" of energy) through a single slit at a time, with a sufficiently long interval in between, and eventually a spot builds up that looks just like the one produced when a very intense (many photons) light was sent through the slit. But then, a strange thing happened. When one sends a single photon at a time (waiting between each laser pulse, for example) toward the screen when both slits are open, rather than two spots eventually building up opposite the two slit openings, what eventually builds up is the interference pattern of alternating bright and dark lines! Hmm... how can this be, if only one photon was sent through the apparatus at a time?

The answer is that each individual photon must, in order to have produced an interference pattern, have gone through both slits! This, the simplest of quantum weirdness experiments, has been the basis of many of the unintuitive interpretations of quantum physics. We can see, perhaps, how physicists might conclude, for example, that a particle of light is not a particle until it is measured at the screen. It turns out that the particle of light is rather a wave before it is measured. But it is not a wave in the ocean-wave sense.

It is not a wave of matter but rather, it turns out, that it is apparently a wave of probability. That is, the elementary particles making up the trees, people, and planets – what we see around us – are apparently just distributions of likelihood until they are measured (that is, measured or observed). So much for the Victorian view of solid matter!

The shock of matter being largely empty space may have been extreme enough. If an atom were the size of a huge cathedral, then the electrons would be dust particles floating around at all distances inside the building, while the nucleus, or center of the atom, would be smaller than a sugar cube. But with quantum physics, even this tenuous result would be superseded by the atom itself not really being anything that exists until it is measured. One might rightly ask, then, what does it mean to measure something? And this brings us to the Uncertainly Principle first discovered by Werner Heisenberg. Dr. Heisenberg wrote, "Some physicists would prefer to come back to the idea of an objective real world whose smallest parts exist objectively in the same sense as stones or trees exist independently of whether we observe them. This however is impossible."

Perhaps that is enough to think about for now. In the next essay, we will examine, in some detail, the uncertainty principle as it relates to what is called "the measurement problem" in quantum physics. We'll find that the uncertainty principle will be the key to performing the double-slit experiment over astronomical distances, and demonstrate that quantum effects are not just microscopic phenomena, but can be extended across the cosmos.

Quantum Astronomy, Part II: The Heisenberg Uncertainty Principle

This is the second article in a series of five, each with a separate explanation of different quantum phenomena. Each article is a piece of a mosaic, so every one is needed to understand the final explanation of the quantum astronomy experiment we propose, possibly using the Allen Array Telescope and the narrow-band radio wave detectors being built by the SETI Institute and the University of California, Berkeley.

In the first article, we discussed the double-slit experiment and how a quantum particle of light (a photon) can be thought of as a wave of probability until it is actually detected. In this article, we will examine another feature of quantum physics that places fundamental constraints on what can actually be measured, a basic property first discovered by Werner Heisenberg, known in its simplest form as the "Heisenberg Uncertainty Principle."

In scientific circles we are perhaps used to thinking of the word "principle" as "order," "certainty" or "a law of the universe." So the term "uncertainty principle" may strike us as something akin to the terms "jumbo shrimp" or "guest host" in the sense of juxtaposing opposites. However, the uncertainty principle is a fundamental property of quantum physics initially discovered through somewhat classical reasoning – a classically-based logic that is still used by many physics teachers to explain the uncertainty principle today. This classical approach is that if one looks at an elementary particle using light to see it, the very act of hitting the particle with light (even just one photon) should knock it out of the way so that one can no longer tell where the particle actually is located – just that it is no longer where it was.

Smaller wavelength light (blue, for example, which is more energetic) imparts more energy to the particle than longer wavelength light (red, for example, which is less energetic). So, using a smaller, more precise "yardstick" of light to measure position means that one "messes up" the possible position of the particle more by "hitting" it with more energy. While his sponsor, Niels Bohr (who successfully argued with Einstein on many of these matters), was on travel, Heisenberg first published his Uncertainty Principle paper using this more-or-less classical reasoning just given. The deviation from classical notion was the idea of light comes in little packets or quantities, known as "quanta," as discussed in article one. However, the uncertainty principle was to turn out to be much more fundamental than even Heisenberg imagined in his first paper.

Momentum is a fundamental concept in physics. It is classically defined as the mass of a particle multiplied by its velocity. We can picture a baseball thrown at us at 100 miles per hour having a similar effect as a bat being thrown at us at ten miles per hour; they would both have about the same momentum, although they have quite different masses. The Heisenberg Uncertainty Principle basically stated that if one starts to know the change in momentum of an elementary particle very well (that is, usually, what the change in a particle's velocity is) then one begins to lose knowledge of the change in the position of the particle, that is, where the particle is actually located. Another way of stating this principle, using relativity in the formulation, turns out to be that one gets another version of the uncertainty principle. This relativistic version states that as one gets to know the energy of an elementary particle very well, one cannot at the same time know (i.e., measure) very accurately at what time it actually had that energy. So, we have in quantum physics, these are called "complementary pairs." (If you'd really like to impress your friends, you can also call them "non-commuting observables.")

We can illustrate the basic results of the uncertainty principle with a not-quite-filled balloon. On one side, we could write "delta-E" to represent our uncertainty in the value of the energy of a particle, and on the other side of the balloon write "delta-t," which stands for our uncertainty in the time the particle had that energy. If we squeeze the delta-E side (constrain the energy so that it fits into our hand, for example), we can see the delta-t side of the balloon get larger. Similarly, if we decide to make the delta-t side fit within our hand, the delta-E side gets larger. But the total value of air in the balloon doesn't change; it just shifts. The total value of air in the balloon in our analogy is called one quantity, or one "quanta," the smallest unit of energy that exists in quantum physics. You can add more quanta-air to the balloon (making all the values larger, both in delta-E and delta-t), but you can never take more than one quanta-air out of the balloon in our analogy. This means "quantum balloons" do not come in packets any smaller than one quanta, or photon.

It is interesting that the term "quantum leap" has come to mean a large change in something, rather than the smallest possible change; in fact, the order of the dictionary definitions of "quantum leap" have been switched, with the popular usage listed first, and the physics usage, which is the opposite, listed second. If you say to your boss, "We've made a quantum leap in progress today," this can still, however, be considered an honest statement of making as small amount of progress as possible in keeping with the laws of physics!

When quantum physics was still young, Albert Einstein and his colleagues would challenge Niels Bohr and his colleagues with many strange quantum puzzles. Some of these included effects that seemed to imply that elementary particles, through quantum effects, could communicate faster than light. Einstein was known to then imply that we really could not be understanding physics correctly for such effects to be allowed to take place because, among other things, such faster-than-light connectedness would deny the speed of light limit set by relativity. Einstein came up with several such self-evidently absurd thought experiments to perform, the most famous being the EPR paradox, which was named after the three authors of the paper, Einstein, Podolsky and Rosen. It showed that faster-than-light communication would appear to be the result of certain quantum experiments, and therefore argued that quantum physics was not complete, and that some factors must be, as yet, undiscovered. This led Niels Bohr and his associates to formulate the "Copenhagen Interpretation" of quantum physics reality. This interpretation (overly simplified, in a nutshell) is that it makes no sense to talk about an elementary particle until it is observed, because it really doesn't exist unless it is observed. In other words, elementary particles might be thought of not just as being made up of forces, but some constituents of it that must be taken into account are the observer or measurer as well, and that the observer can never really be separated from the observation.

Using the wave equations formulated for quantum particles by Erwin Schrodinger[1], Max Born was the first to make the suggestion that these elementary particle waves were not made up of anything but probabilities! So the constituents of everything we see are made up of what one might call "tendencies to exist," which are made into particles by adding the essential ingredient of "looking." Looking, as an ingredient itself, it must be noted, took some getting used to! There were other possible interpretations, but none of them were consistent with any sort of objective reality as Victorian physics had known it before. The wildest theories could fit the data equally well, but none of them allowed the particles making up the universe to consist of anything without either an underlying faster-than-light communication (theory of David Bohm), another parallel universe branching off ours every time there is a minute decision to be made ("many worlds" interpretation), or the "old" favorite, the observer creates the reality when he looks (the Copenhagen Interpretation).

Inspired by all these theories, John Bell, a physicist at CERN in Switzerland, came up with an experiment in 1964 that could perhaps test some of these theories and certainly test how far quantum physics was from classical physics. By then, quantum physics was old enough to have distinguished itself from all previous physics to the point that physics before 1900 was dubbed "classical physics," and physics discovered after 1900 (mainly quantum physics) was dubbed "modern physics." So, in a sense, the history of science in broken up into the first 46 centuries; Starting with Imhotep, who built the first pyramid as the first historical scientist, to the end of the classical period in 1900, and then from the beginning of the 20th century forward, with quantum physics. You can see that the age of modern physics, with this new fundamental view of science, is quite young. It might even be fair to say that most people are not even aware, even after a century, of the great changes taking place in the fundamental basis of scientific endeavors and interpretations of reality.

Bell proposed an experiment that could measure if a given elementary particle could "communicate" with another elementary particle farther away faster than any light could have traveled between them. In 1984, a team led by Alain Aspect in Paris did this experiment and indeed this was undeniably the apparent result. The experiment had to do with polarized light. To illustrate, let's say you have a container of light, and the light is waving all over the place and – if the container is coated with a reflective substance, except for the ends – the light is bouncing off the walls. Picture a can of spaghetti with noodles in all orientations as the different directions of the random light waves. Then, we'll place polarizing filters at the ends of the can. Now, only light with a given orientation (like the noodles that are oriented up and down) can get out, while back and forth light waves (the noodles that are not oriented this way) cannot get out. If we rotate the polarizers at both ends by 90 degrees, we would then let out back and forth light waves, but not up and down light. If we rotate the ends, so that they're at an angle of 30 degrees to each other, about half of the total light can get out of the container – one-fourth from one side of the can, and one-fourth through the other side.

This is similar to what Bell proposed and Aspect demonstrated using a long tube. When their tube was rotated at one end, creating a 30-degree angle with the other side so that only half the light could escape, a surprising thing happened. Before any light could have had time to travel from the rotated side of the tube to the other side, the light coming out of the opposite side from the one that was rotated changed to one-fourth instantaneously (or as close to instantaneous as anyone could measure). Somehow, that side of the tube had gotten the message that the other side had been rotated faster than the speed of light. Since then, this experiment has been confirmed many times.

Bell's formulation of the fundamental ideas in this experiment have been called "Bell's Theorem," and can be stated most succinctly in his own words: "Reality is non-local." In other words, not only do the elementary particles that make up the things we see around us not exist until they are observed (Copenhagen Interpretation), but they are not, at the most essential level, even identifiably separable from other such particles arbitrarily far away. John Muir, the 19th century naturalist, once said, "When we try to pick out anything by itself, we find it hitched to everything else in the universe." Well, he might have been surprised how literally – in physics as well as in ecology – this turned out to be true.

In the next, article, we'll combine the uncertainty principle with the results of Bell's Theorem, and increase the scale of the double slit experiment to cosmic proportions with what Einstein's colleague, John Wheeler, called "The Participatory Universe." This will involve juggling what is knowable and what is unknowable in the universe at the same time.

---

[1] This is the same Erwin Schrodinger who came up with a thought experiment regarding a cat in a box being both alive and dead as a result of quantum uncertainty.

For more information:

Werner Heisenberg: http://www.aip.org/history/heisenberg/

Niels Bohr: http://en.wikipedia.org/wiki/Niels_Bohr

Albert Einstein: http://scienceworld.wolfram.com/biography/Einstein.html

John Bell: http://physicsweb.org/articles/world/11/12/8

Erwin Schrodinger: http://scienceworld.wolfram.com/biography/Schroedinger.html

Max Born: http://en.wikipedia.org/wiki/Max_Born

Quantum Astronomy, Part III: Knowability and Unknowability in the Universe

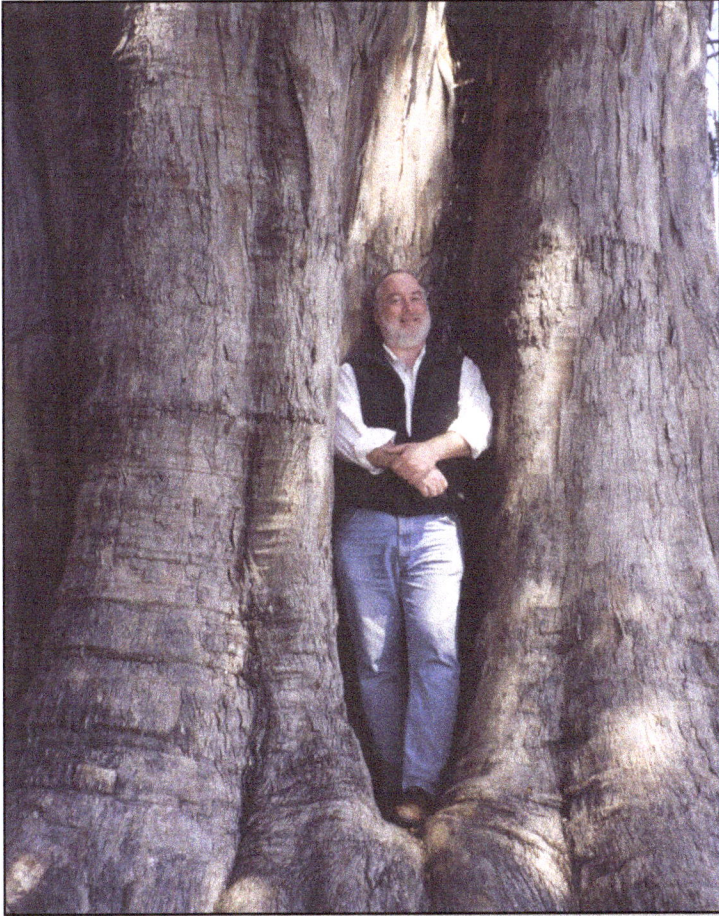

In the previous two articles we discussed the basic double-slit experiment that demonstrates the dual nature of light – wave and particle – and then the Heisenberg Uncertainty Principle, which demonstrates the complementary (mutual exclusion) of what can be measured at the same time. In this article, we'll discuss the more basic interpretation of quantum physics in terms of what can be known or not known, and how this affects the measurement results.

John Bell formulated the uncertainty principle in terms of what one could know or not know in an experiment. Several Bell-type experiments have successfully shown that this would seem to be the simplest interpretation of the situation. Taking our double-slit example, if one puts a detector at one or the other of the two slits – even one that does not destroy the photon, electron or whatever particle as it goes through the slit – then an interference pattern does not appear at the detector. This is because one has set up an experiment in which one can "know" which path the particle took, i.e., which slit the particle went through. As long as one can tell this, then the particle cannot go through both slits at once, and one no longer gets an interference pattern.

What if we decide not to look at the detector set up next to one or the other of the slits? Well, we still won't get an interference pattern, because the potential exists to be able to tell which path the photon, for example, traveled. Even this potential (i.e., "knowability") is enough to stop the formation of an interference pattern. All ability to detect which path an elementary particle took (in this case a photon of light) must be removed to obtain interference. In other words, we must not even be able to tell – even in principle – which path the elementary particle took in order for it to "take" both paths and form an interference pattern.

This is the most fundamental concept in quantum physics – knowable and unknowable. It's from this fundamental concept of the uncertainty principle that we'll approach our quantum astronomy experiment. It has been experimentally verified that if we can know which path a photon traveled, then an interference pattern is not possible. But if we can become ignorant of which path the photon (or any elementary particle) took, then an interference pattern is assured. That is, if one is ignorant of which path the photon took, the interference pattern is not just possible, it must occur.

This last point can produce some decidedly non-classical effects. One example is the phenomenon known as "quantum beats." Picture an atom (classically, for now) as consisting of a nucleus with electrons jumping all around it. Electrons do not move smoothly away from and toward their central nucleus; they take discrete steps (energy quanta, actually) to transition from a lower to a higher (farther from the nucleus) orbital level. They actually disappear from one level and reappear at another, but are never found in between the two. As an electron "jumps" from a higher level step to a lower level step, it emits a photon of light. Just the fact that we cannot, even in principle, tell which energy level jump the electron took, is enough to produce a special kind of interference fringe called "quantum beats."

While picturing probability distributions as classical waves (like water waves) may be helpful for beginning physics students, real quantum wave phenomena are decidedly non-classical, and produce decidedly non-classical results. They are not waves made of anything but probabilities, or tendencies to exist. Yet they can interfere with each other in a wave phenomena-type way before they are measured, and so "turn" from probability waves to measured particles. This "collapse of the wave function" is also said to take place instantaneously, as we'll discuss further in article four.

Einstein wrote several times, "God does not play dice with the universe." Quantum physics, however, has reduced everything to probabilities mathematically, and such a formulation inherently implies dice rolling for all possibilities until a measurement is made. Richard Feynman pointed out that the mathematics really does mean all possibilities. Every elementary particle takes every path it possibly can – a kind of infinite-slit experiment – and then these infinite numbers of paths all cancel in the multi-dimensional mathematics – called Hilbert space – so that only one result is finally measured.

However, a colleague of Einstein's, Professor John Wheeler of Princeton University, has pointed out that one could take another interpretation, an interpretation he has dubbed, "The Participatory Universe." In this approach, we can look at the universe as directly participating in each quantum effect in real time. In other words, the concept of a First Cause starting things off (winding up the clock of the universe, you might say) and then leaving the laws of physics to run things, may be what is incorrect in the basic approach of classical physics. Rather, in this participatory scenario, the Cause of the laws of physics remains an active Participant. (If one would like to also draw some religious points into such discussions I would just say that it is important to understand what is being said and what is not being said here – that is, to not oversimplify what went into Wheeler's introduction of this interesting proposed conceptualization for quantum reality.)

Wheeler came up with a Gedanken experiment (i.e., a thought experiment) that he called the "delayed choice." experiment. He proposed a huge scaling up (to cosmic proportions) of Young's double slit experiment that we've talked so much about. In this Gedanken experiment, gravitational lenses, which can bend light from distance quasars or galaxies, are used as if they were giant slits, to create two paths for photons from a quasar or distance galaxy. General Relativity shows that masses in space can bend light.

The first support for the confirmation of Einstein's theory of relativity came with the measurement of the bending of light from stars by the Sun as they passed close behind it during a total solar eclipse. Light was indeed bent by the mass of the Sun (that is to say that space-time was curved near large masses). It turns out that large masses, like galaxies, that are rather close to being directly in between a distant quasar-galaxy and us, will bend the light from the distant object toward us. We can think of light coming toward us more or less directly from a distant quasar-galaxy (let's call this path A) while light shining from this quasar also heads off into space at a slightly different angle.

This light, however, encounters a massive galaxy along the way, so the light rays that would normally have missed the Earth get bent towards us as well (we'll call this light path B). Because of this, it appears as if we have two quasars with a massive galaxy in between. However, this is just one quasar whose light rays are coming more or less directly toward us along path A, and whose second image appears on the other side of the massive galaxy image, with this latter image being the rays traveling along bent path B (i.e., bent toward us). That's why it appears we have two quasars when we actually have two images of the same quasar.

Wheeler realized these two paths constituted a kind of double-slit experiment, where the slits were the two gravitational lens images. The two paths of light from the quasar might be used then to interfere with each other. However, this could be done – according to Bell's approach to the uncertainty principle – only if we could not tell which path any particular photon had traveled. One way of being able to avoid knowing which path an individual photon took is to make the paths equal (within the uncertainty principle) so we could not tell whether any photons arriving had traveled along path A or path B. Even if there were a flare in the quasar, the flare (peak in brightness) would arrive at the same time at Earth, and so we could not use timing to tell which way it came. (We can see that if the paths are not equal, the light from the flare would arrive along path A before path B, and so we could tell the path difference, which would negate the possibility of getting an interference pattern.)

Wheeler "solved" this problem by adding an immensely long fiber optics cable to path A to make it as long as path B. The fiber optics cable turned out to have to be over a light year long in this case, so it really was truly a Gedanken experiment without much hope of realization – but I will propose a possible solution to this problem in the next article. The delayed choice part of the experiment was, nevertheless, still very interesting. Given, then, that we can achieve an interference pattern in this way, we should be able to put a detector at the intersection of light paths A and B and just re-do a cosmic-scale version of Young's double-slit experiment, where light-photons from quasar images A and B crossing the universe are equivalent to light going through slits 1 and 2 in the laboratory.

One of the founders of quantum physics, P.A.M. Dirac, noted that, at least in Young's double-slit experiment, an interference pattern could only be found if each photon only ever interfered with itself – that is, each single photon had to go through both slits and so only interfere with itself, and not with any other photon. This certainly made sense in terms of the interference experiment being done with one photon at a time, and still producing interference. There are also conservation of energy arguments – two photons should not be expected to produce 4 times the energy when they meet sometimes – making a bright line – and no energy when they meet at other times – making a dark line in the interference pattern. If we performed the Wheeler delayed choice experiment with single photons being detected at a time, we would still expect – if we could not tell which path, A or B, the photons traveled along – that an interference pattern would result where the two paths met. However, if we moved the detector to just detect photons along path A, then these photons will have just traveled along path A (by classical reasoning). Or, similarly, if we move the detector to intersect path B photons, then the photons will have traveled only path B.

The interesting part of Wheeler's thought experiment is that the quasar emitting the photons is about one billion light years away – that is, the light from this quasar is supposed to have taken a billion years to travel to Earth. It seems perplexing that any given photon will have had to have traveled both paths when you put the detector at the intersection of both paths, but then one path or the other path when you decide to put the detector directly into one of these paths, rather than at their intersection.

In other words, how can your decision as to where to put the detector affect the path of a given photon a billion years after it supposedly started along one of the paths toward Earth – long before humans even existed on this planet, much less discovered quantum physics? It would appear that what has "happened" in the distant past in this case may be determined by what is happening right now, even though it is supposed to have "happened" over a billion years ago. The choice of which path, in other words, has somehow been "delayed." One might view this as the Universe playing more the part of an active participant in what is happening rather than just in what has happened in the past in this case. This is known as the "Participatory Universe" conceptualization.

This interesting Gedanken experiment points out what may be the main difference between general relativity and quantum physics. In general relativity, time is a definite dimension, part of the already unalterable space-time continuum. While in quantum physics, time is, at best, a variable, and is also quantized (i.e. there are particles of time). Far from being an absolute, time in quantum physics is a not a solid background upon which particles in space change. In quantum physics, time is not yet, in a sense, really even there until the "time particles" are measured.

In our fourth article, we will talk about the possible realization of Wheeler's Gedanken
experiment, which may open up a whole new field of investigation – a field we'll call "quantum astronomy."

Quantum Astronomy, Part IV: A Cosmic-Scale Double-Slit Experiment

This is the fourth article in a series of five, each explaining different quantum phenomena. Each of these is a piece of a mosaic, so every one is needed to understand the final explanation of the quantum astronomy experiment we're proposing, possibly using the Allen Telescope Array and the narrow-band radio wave detectors being built by the SETI Institute and the University of California, Berkeley.

In the preceding three articles, we discussed Young's double-slit experiment, where light was shown to behave as a wave. We also discussed the birth of quantum physics, where light was also shown to behave like a particle. In the second article, we discussed a basic limitation on measurement imposed by the Heisenberg Uncertainty Principle and how one may "trade" knowledge of one measurement for another. In article three, we discussed John Bell's concept of knowability and unknowability, and John Wheeler's Gedanken (thought) experiment creating a cosmic-scale, double-slit experiment requiring an immensely (billions of miles) long fiber optics cable.

It is the application of Bell's concept of knowability and unknowability that we shall now apply to the uncertainty principle in order to try to perform Wheeler's cosmic double-slit experiment over cosmic distances that we shall discuss in this article.

In order to realize this experiment, however, one must come up with a substitute for this unbuildable long fiber optics cable, and this is where the SETI Institute's new Allen Telescope Array and its narrow-band radio wave detectors can play an important part. SETI radio projects use the fact that, as far as we know, no natural (i.e., non-technological) source of radio waves can make a very narrow-band radio channel. When you tune to a station on the radio, one turn and you are on another channel. If you tune to a radio galaxy, however, you can turn the dial many dozens of times and you will still be on the same channel, so to speak – you will hear the same sounds. In other words, as far as we know, only technology can make a narrow (1 Hertz wide) radio channel. Thus, looking for narrow-band signals in space should be a good way to look for evidence of any radio-technological civilizations around other stars. Fortunately for quantum astronomy, it also turns out that an extremely narrow-band radio channel can also be used to replace that unrealistically long fiber optics cable! But to explain just how this can be done, we need to first look again at the uncertainty principle.

When a colleague, Dr. David P. Carico of San Francisco State University, and I began thinking about actually carrying out Professor Wheeler's delayed choice experiment, we realized that the uncertainty principle needed to be satisfied in order to obtain an interference pattern. That is, to be ignorant of which path the light traveled – along path A (directly from the quasar), or along path B (the path most bent by the gravity of the intervening galaxy back toward Earth), so that it could "travel both paths" and so interfere with itself. The terms "travel" and "path" as applied to a photon-wave, of course, do not have any real meaning in quantum physics if the particle-nature does not exist until it is measured. But for now, we will use such terms, as it is difficult to speak of quantum effects without some reference to our classical notions of space and time. The energy-time uncertainty principle referred to the fact that knowing the energy of a given particle meant that we could not know precisely the time the particle had that energy. And, "complementarily" (the term for this that Niels Bohr used), if one knows the time to a high precision, we cannot then know what that energy was with greater accuracy than the basic quantum value. This quantum value is called "Planck's constant," or one quantum of energy, and is actually quite a small value, so we do not usually notice this uncertainty constraint in everyday activities.

Now, in thinking about how to do this experiment, we thought that perhaps it might be possible to "trade" knowledge of energy for knowledge of time, but in this case the time would be the delay time between the two paths of the gravitational lens images, A and B. The uncertainty in energy then might be able to replace the hugely long fiber optics cable with, instead, a very narrow-band radio detector. We have seen that we can trade knowledge of energy for knowledge of time (recall our balloon image with "delta-E" written on one end and "delta-t" written on the other).

We also recall that if we can tell which path each photon traveled, we will not get an interference pattern, but rather just a picture of a quasar at A and another (image of it) at B. To understand this "trade" then, let's take just a bit closer look at what we mean by a narrow-band radio wave.

It is known in the physics of electromagnetic waves that longer waves have less energy than shorter waves. The blue light we see has more energy per photon than the red light we see. This can be extended to lower energy infrared photons, and higher energy ultraviolet photons, and to very low energy radio photons and even to very much higher energy x-ray photons. In photography, using a filter on the camera lens can allow only blue light or red light into the camera. Sunlight is usually a whole mixture of blues, greens, yellows, oranges, reds and so on, and therefore also a mixture of photons of light of all kinds of energy, high and low. When we use a red filter, for example, we are cutting out the higher energy blue photons from going into the camera, and so only detects the lower energy red light. The narrower the filter, the less range of energy is let into the camera.

Similarly for radio detectors, if we have a broadband detector, we are letting in radio waves of all sorts of energies all at once. However if we have a very narrow-band radio detector (such as are used in the search for extraterrestrial intelligent technology), we are highly constrained in the range of energies being detected. Only the radio photons of a very narrow spread in energy are actually measured. Remembering the uncertainty principle for energy and time, we can recognize that narrow-band radio detectors represent a constraint on the value of the energy being measured. Now what about time, however? For that, let's look at the crossing of the radio waves (which is just long wavelength light) coming along paths A and/or B. We can only get an interference pattern if we cannot tell (or even potentially be able to tell) which path a radio photon took to reach our detector. But, if the difference in the travel time between paths A and B is long enough (this is called the "delay time" of the gravitational lens), then there is plenty of time to detect, for example, if a flare went off at the quasar so that image A brightened, followed by image B some time later (the delay time). This is actually how the delay time between gravitational lens paths is measured. Now, the next sentence is the most important. If we use a narrow enough radio bandpass, we can potentially constrain the energy to such a precise value that the time uncertainty is so large, it exceeds the actual delay time of the gravitational lens. In other words, we can constrain the energy (by using narrow-band radio detectors) so much that we can exceed the ability – even potentially – of measuring which path the photon travels, because our uncertainty in the arrival time of the photon is now larger (because of the uncertainty principle) than the actual delay time or travel time difference between paths A and B. Therefore, we cannot tell which path the photon traveled along, and so should get an interference pattern at the detectors. A very narrow-band (but real) radio detector then, can substitute for an unrealistically long fiber optics cable to get an interference pattern at the intersection of paths A and B.

So, how do we proceed to do this? We can start observing the gravitational lens using a radio telescope with very narrow-band detectors. We set the detectors on the narrowest-band possible (let's say one one-hundredth of a Hertz, which means we know the wavelength – and therefore energy – of the radio wave coming in to within one one-hundredth of a wavelength per second.) We focus the two images of the quasar across each other, and, if the delay time is not too long – not longer than 100 seconds in this case – we will obtain an interference pattern. This means we cannot know which path the radio photons "took." For simplicity, we also assume there are no detectable rapid fluctuations from the quasar, although there are ways of dealing with this effect as well, using "choppers" in the path of the incoming light. Now, what happens if we increase the allowable energies being detected (i.e., increase the bandpass of the radio detectors)? At first, we may still get an interference pattern. But if we continue to increase the bandpass, at some point the interference pattern will disappear, and we shall simply get a (radio) picture of a quasar at location A, and another of its image at location B. The interference pattern will have disappeared at exactly the point where we could begin to tell which path the photons took. In other words, by allowing ourselves to become more and more ignorant of the energy of the radio waves arriving, we simultaneously allowed an increased knowledge (according to the uncertainty principle) of the time interval. And when we decreased our knowledge of the energy to the point where our knowledge of the time interval could drop below the actual delay time between light paths of the gravitational lens, we could, at least in principle, tell which path each photon took. Thus, the uncertainty principle "kicks in" and says that we cannot know which path a photon took and still get a wave phenomenon (i.e., an interference pattern). We cannot have one's photon and wave it too.

Therefore, we may be able to use very narrow-band radio detectors to realize the delayed choice (perhaps no longer just Gedanken) experiment proposed by Wheeler. What is of interest in doing such an experiment? First, it may represent a possible way to directly measure delay times for gravitational lenses that don't vary much in brightness, and such delay times can be used to measure the expansion rate of the universe directly (this parameter is called the "Hubble constant"). But more intriguing, perhaps, is that it can possibly provide a measure of the minimum time it takes for a wave to "become" a particle. If the quasar is one billion light years away (that's about six billion trillion miles), and the interference pattern is being formed by a probability wave that is traveling along both paths A and B, then when we increase the bandpass (say, over one hour's time) to the point where the wave becomes a particle (photon), then we might be able to speak in terms of the wave "becoming" a particle at the minimum rate of a billion light years per hour. This rate is considered in most quantum physics formulations to be instantaneous, but one is reminded of Galileo and a colleague standing on opposite hillsides with lamps trying to measure the speed of light. When one opened the lampshade, as soon as the other saw it, they opened their lampshade, and so, back and forth. They decided that the speed of light was either instantaneous or very, very fast. It turned out to be very, very fast (186,300 miles per second) – far too fast to measure with shaded lamps on nearby hills. So, perhaps quantum astronomy may someday allow such a measurement of the speed of the wave-to-particle transition, if it is not instantaneous. What we have outlined here is just one experiment of many possible experiments that could be performed in what may be one of the most interesting new fields of the 21st century, quantum astronomy.

Quantum Astronomy, Part V: Information in the Universe

No field has given me as much feedback as the series on quantum astronomy. I think people intuitively believe that quantum physics is still redefining how we think of science and what we think the fundamental nature of reality may be, and enjoy participating in this amazing modern adventure.

To summarize the preceding articles on quantum astronomy, in the first article, we looked at the double-slit experiment and how it appears to indicate that a single particle of light (a photon) travels through two slits (apertures) to make an interference pattern, apparently being in two places at once, and yet still be detected as a small particle when it registers on a detector screen. In the second article, we looked at Heisenberg's Uncertainty Principle, which requires that certain pairs of measurable quantities (position and momentum, for example) cannot both be measured accurately simultaneously. Time and energy are another such set of "complementary pairs," so that if one measures the energy of a particle really well, one cannot tell very accurately at what time the particle had that energy. This uncertainty principle can be manipulated, you might say that you can trade off one kind of information for another, as long as ignorance is conserved.

In the third article, we noted that waves associated with particles in quantum physics are waves of probability (not waves like ocean waves, although they do share many characteristics). So what we can know or cannot know about, for example, which path a photon took to a detector, actually determines what we will detect whether an interference pattern is detected or not. If we cannot tell which path a photon took to a detector, we can get interference, but not otherwise. And finally, in the fourth article, we discussed doing a cosmic-scale double-slit experiment, first proposed by Albert Einstein's colleague, John Wheeler of Princeton University, where a decision about which path a photon takes around a gravitational lens (a galaxy aligned so it can bend light from a more distance quasar) can be decided long after even billions of years, after the photon had supposedly already left the source and traveled along one path or the other. This was called the "cosmic-scale delayed choice" experiment.

To review this experiment, Wheeler proposed that light from a quasar about 7 billion light years away is split by a gravitational lens, and so we have light traveling to us along two paths: A, the shorter path, and B, the longer path, that encounters more of the gravitational lensing galaxy and whose path is "bent" toward us. If a fiber optics cable (trillions of miles long would be needed) could be used to make the distance along the shorter path, A, equal to the distance along path B, then we could get an interference pattern rather than just an image of A superimposed on an image of B at the detector. But interestingly, at the detection rate of one photon at a time, that would mean we could decide to have the photon travel both paths at the last moment rather than just path A or B, and we'd be deciding this 7 billion years after the photon supposedly left the quasar! That's why this experiment really meant delayed choice, to the point where Wheeler could talk about his hypothetical experiment in terms of altering "history." But it could only be thought at the time (such thought-only experiments were dubbed *"gedanken"* experiments by Einstein).

Changing this experiment from a *gedanken* experiment to a performable experiment, my colleague, Dr. David Carico, and I proposed that we might actually utilize the uncertainty principle itself to replace the trillions-of-miles-long fiber optics cable. This notion was based on the idea that since knowability or unknowability is the important consideration (rather than the actual distances involved), we proposed not so much to make the two paths which a photon traveled equal, but rather to just render any difference in the length of the two paths unmeasureable (i.e., unknowable). We proposed that by knowing the energy of the photon very well (by using a narrow-band radio filter, for example) that the time that the photon actually had that energy would be unknowable (since time is the complimentary pair of energy). So, if the unknowability in the time is unmeasureably longer than the delay time between the light paths of the gravitational lens itself, then the two paths are, essentially, "unmeasureably equal," and one cannot tell which path the photon took. If one persists in thinking classically, the photon can then be said to have taken both paths. To put it in physics-ese, we have used the uncertainty principle as a quantum eraser; it erases the quantum nature of a photon, making it a probability wave again, which can exist (if probability wave can be said to exist) along both possible paths again.

We did have to go through some mighty refereeing to get this paper in print, however. One of the biggest doubts about this experiment working was related to using it on extended objects in the sky. It was said that one may measure a point source traveling along two paths, but what if the source is a whole extended galaxy? Well, even galaxies can be thought of as being made up of a lot of point sources, so we argued that the technique would still nevertheless apply, as long as one could not tell what the extent of the actual galaxy (angular size on the sky) was. We did this by introducing what is called a Mach-Zehnder Interferometer (MZI) which, unlike a double-slit setup, cannot tell the angular extent of a photon source because it does not produce an interference pattern. It only indicates whether interference is taking place or not. For those familiar with the MZI, the gravitational lens itself is the first beam splitter in the system and has an effective refractive index so you can change the phase of the light. For those of you not familiar with the MZI, thanks for hanging in there so far!

We also talked with many physicists about this idea and all were encouraging. Freeman Dyson of the Princeton Institute for Advanced Studies told us, "I think you're OK." Andre Linde of Stanford University said, "These things are tricky." Daniel Greenberger of City College of New York said, "I think it is worth a try." And Wheeler, at a scientific meeting, on the occasion of his 90th birthday, said, "That's very interesting. I hope you succeed." Of course, the actual referees for our paper were more detailed, and the process did drag on for a couple of years. Scientists are usually very friendly and happy to discuss new ideas, but when something is going into the refereed scientific literature, that is another story.

One referee wrote, "The validity of the claim that interference would be observed between extended sources if observed through a sufficiently narrow-band filter is absolutely critical." If it is right, the implications would be extremely profound, and extend far beyond the narrow confines of measuring time delays in lensed systems, as it would completely undermine the conventional understanding of how interferometry works. I have to confess that as I read this, at this point I thought, "Gulp!" But I also realized that barring anything we and the referees and editors overlooked, on the other hand, if this experiment did not work it would be a more radical departure for physics than if it did. This is because it would imply that the quantum uncertainty principle itself did not apply in some circumstances or did not, for example, extend over macroscopic distances. So, with this argument, our paper was finally accepted.

The great quantum physicist, Richard Feynman, once said (to paraphrase): "If you think you understand quantum physics then you don't understand enough to understand that you don't understand it!" And Einstein himself once wrote, "I have thought a hundred times as much about the quantum problems as I have about general relativity theory." We can relate. And you are also most welcome to join Einstein's 'hundred times' club. You, too, may begin thinking of the universe, not so much in terms of material objects, but rather in terms of information. And as quantum measurement begins to leave the laboratory and extend throughout space, I think we're all in for a lot of surprises. And a lot of fun, too.

Sherlock Holmes and The Case of the Vanishing Robbers

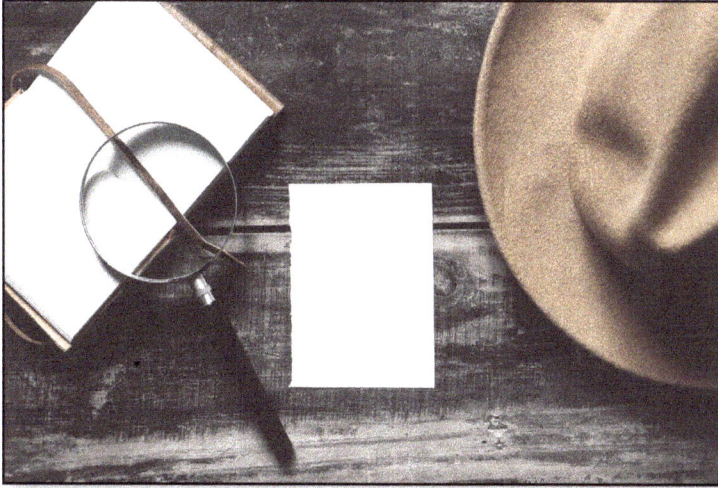

Author's note: I once read that Arthur Conan Doyle's character, Sherlock Holmes, did not care about Earth's orbit, so I thought it might be a good idea to have him collaborate with the Royal Greenwich Observatory. Being in both the "Doyle" and "Astronomy" clans, I thought I'd give it a go. So, here is "The Case of the Vanishing Robbers."

When writing about cases involving my remarkable friend, it has not always been the most profound cases that I have endeavored to portray, but more often the cases that have shown that insight and attention to detail he so remarkably displayed. However, some cases have been so singular that, although they may have been easy for my friend, they possessed a novelty about them that certainly makes them worthy of note. Such was the case of the vanishing robbers.

It was early evening at our home at 221 B Baker Street, and Holmes was approaching the state of gloom that he always descended to when challenging cases were few and far between. There had been nothing of note in the crime pages of the papers for several days, and a clear cold snap had descended on London, tending to make even a very brisk mid-evening walk outside a bit uncomfortable.

"Holmes," I directed, "see what you can tell of this fellow from a small photograph I have," I said in a feeble attempt to keep his spirits up. I knew the fellow to have been a close cousin of my wife's, and she had had the picture taken a number of years ago.

"Well, let's have a look," said Holmes in an attempt at good nature, and he had hardly glanced at the picture when suddenly, Mrs. Hudson entered the room. "Telegram for you Mr. Holmes," said she, placing it in his left hand as he put my picture down on the table. "Watson, it is Lestrade. We must go at once! Can you accompany me on this adventure? I shall be grateful for your assistance."

"I wouldn't miss it for anything," I replied. "What has occurred, Holmes?"

"The London First National has been robbed this evening, and the events of the robbers' escape appears to have been most singular."

I picked up my picture, got my coat, and we soon found ourselves in a cab on the way to Market Square. "One should never cloud one's deduction by the formulation of scenarios without information, Watson. I have, however, often found it difficult to refrain. It is now that I will appreciate your little attempt at diversion."

"Surely you had only a glance, Holmes. I certainly would not have expected you to tell much."

"True enough, Watson. I could only tell that the man was a close relative of your wife's...probably a cousin, was a student at the time of the photo – which incidentally was probably taken over a decade ago by the colour of the paper – he was left-handed, has passed some time in the Orient, and had certainly been a member of the crew team."

"By Jove, Holmes! I apologize for my feeble attempt at a challenge, but how on Earth did you know, for example, that George [for that was my cousin's name] had been in the Orient for an extended time?"

"Surely it is obvious, Watson."

"It is certainly not obvious to me, I assure you," said I, drawing the picture from my coat pocket.

"George, then, has spent enough time in the Orient to have lost at least one button to his shirt, thereby necessitating the replacement of all of them by the ones you see in the picture. They are of a distinctly Oriental origin, and not made of the usual brass or wood found in European circles."

"But how about the rest? The left hand, the crewing, or that he was a student at the time?"

"Clearly a man with his watch bob on the left is left-handed. This is only emphasized by the small callous on the left middle finger, indicating that he did much writing. This drew my attention to his hands which are calloused in that singular manner that rowing produces. Now how can a man do much writing and still have calloused hands unless he is a student both taking classes as well as being a member of the crew team?"

"Amazing, Holmes! And how did you know that he was my cousin?"

"A bit of a guess? Really not much of one given that he was

wearing on his pocket scarf what certainly appears to be the monogram of your wife's family. You had never mentioned that you had a brother-in-law of this age, so that I assumed it must have been a close relative. I'm afraid that I could not tell more in the time I had to see the picture. At any rate, we now seem to be at our destination."

We pulled up to the London First National Bank where various policemen and members of Scotland Yard could be seen swarming all over the area. "I'm sorry, Sir. No one is allowed on these premises. There's been a robbery," a tall constable said as he stepped in front of us.

"Holmes! Glad to see you!" came a voice from the steps of the bank.

"Ah! Lestrade! What is afoot tonight? And what is this about vanishing robbers?"

Stepping aside, the constable smiled and tipped his hat. "As sure as Scotland Yard, Holmes, I can't find the sense of this one. We have been searching the grounds for footprints, which one might expect to be obvious after this afternoon's rain. Indeed, the footprints into the bank are sure. Two men, right here," he said, leading us to a large broken side window down a little alley next to the bank. "But for the life of me, I can't see any coming out. Surely the alarm went off and the place was surrounded within minutes. Something else peculiar too."

"Firecrackers!" said Holmes.

"Why, yes!" replied Lestrade, a little surprised. "But how..."

"Surely you can smell the slight odor of gunpowder. And those little scraps of colored paper over there."

"Well, sure enough. When we arrived, immediately a full two minutes of firecrackers were set off down that little alley. Like to make us nervous wrecks. But the man who set them off was long gone. His tracks got nowhere near the bank window though, so I guess it might have been a coincidence."

"We don't often believe in coincidences, do we, Watson?" said Holmes as he followed the tracks down the alley where they abruptly came to an end in the next street. "He fled in a small trap, here where the wheel tracks are then hopelessly mixed," said Holmes. "Let us return to the main tracks again."

"Lestrade. Were there any witnesses?"

"None, Holmes, I'm afraid. But quite a few people gathered after the firecrackers went off."

Holmes nodded and taking out his magnifying glass, started to examine the two sets of tracks, around which was the broken glass from the windows. "Most interesting...indeed most interesting," he said. "Watson, come here a minute. You are a bit heavier than I. Come and place your foot into the mud here. Ah, as I thought."

"What is it, Holmes?"

"Look at the step size of these prints. It is clear that they were carrying something heavy from the spacing of the steps. Yet their tracks don't really sink in very far into the mud. The ground could hardly have dried since this evening's earlier rain."

"The men must have been fairly lightweight, eh, Holmes?"

"Watson, you are certainly progressing. But there doesn't seem to be anything singular about their footwear." Stooping over, he picked up some small objects next to the footprints. "What do you make of these, Watson?" he said showing me some small little reed-like objects covered with mud.

"I say, Holmes, this may give us a clue to the area these men came from, eh? These little reeds may have been tracked up from some other area."

"Hardly, Watson. You see where I have picked them up from besides the footprints. If they were from the muddy soles they would have lain in the footprints themselves. Ah, and look here!" At the left of two sets of footprints nearest the bank, there seemed to be a small-hatched mark in a corner-like shape, placed well down into the mud.

"Excellent," said Holmes, "it appears that they placed their burden down here, before lifting it up through the window into the bank."

Moving into the bank, Holmes began his fine survey of the premises. In a few minutes, he returned. "They cleaned their shoes well before walking around the bank. Standard tools used in breaking into the vault. No fingerprints...and I suspect that the threads from their gloves I've collected will be a standard type, as well. These men were very careful, yet they must have left more behind. Here are some more of those little straw reeds found in several locations between here and the vault."

"Clearly, they have tunneled out!" said Lestrade with a look of sudden determination on his face. "We are going to find a tunnel somewhere in here, Mr. Holmes. No other way of escape is possible!"

"I think it is a good idea to look for a tunnel, Lestrade," returned Holmes.

"Surely you can't take that seriously, Holmes," I whispered. "No one could have had time to dig a tunnel clear to the other side of the street. We saw no evidence of an exit tunnel around the bank. The sewer line is quite far away..."

"Aha! Here it is!" came Lestrade's voice. Dispatching to the other side of the room, which we had not yet had the opportunity to examine, we saw Lestrade and two constables lifting up several pieces of the floor that had apparently recently been removed. "Well, let's see where this tunnel goes, shall we?"

"It goes nowhere, Lestrade," said Holmes, "nevertheless, it is useful."

"Why, it's just a hole," said Lestrade with a disappointed look. "But I say, what are these?" The same footprints had apparently been in the hole, but what Lestrade held up was most peculiar. He took out three long metal bottles.

"I should think that they are high pressure bottles of some sort," Holmes said. "It is certainly of interest that the robbers would want to bury them instead of either taking them along or just leaving them around. Let me make a quick examination of the hole, and then I think our work here is done, Watson." After a quick look, we bid Lestrade goodbye and found ourselves on the street flagging down a trap.

"What do you make of it?" I asked my friend, expecting him to be getting ready for a half-night's smoking and thinking on the case, as was his usual routine in the face of such a perplexing case.

"Oh, the case is all but solved," he said nonchalantly.

"Surely you haven't solved it already!" said I, incredulous.

"Watson, as I have said, whenever all other possibilities have been ruled out, the improbable, however unlikely, must be the truth. I must say that this case is surely singular in all my experience, however."

"But how did the men escape without being noticed? Could they have retraced their footsteps? And the box? And the funny crisscross prints in the mud? And the hole with the metal bottles in it?"

"Now Watson, you know how I dislike to give away the answer before it is time. You have always, I assume, enjoyed, or at least tolerated, my flair for the dramatic up to now. Patience, and you will come to hear it all. But I will tell you this. By the way these men tried to cover their escape, we know that they will undoubtedly try another robbery soon." With that we flagged down a carriage.

"Coachman," said Holmes, "to the Royal Greenwich Observatory. They will certainly be up on such a clear night."

"What?" I said, more than mildly surprised. "I thought you had no interest in astronomy? You once said that it didn't matter to you whether the Sun went around the Earth, or the Earth around the Sun."

"I may have been premature, Watson. At any rate, it is never too late to learn something new, eh?" he said with a grin. The complicated whims of this man never ceased to surprise me.

The ride to the observatory was cold, given the humidity left from that early evening's rain. However, it had been clear now for some time since, and it was nearing 1 a.m. when we arrived at the door of the observatory. We stepped out, and the stars seemed dazzling on this little London hill. Holmes knocked loudly at the door of the tall silver dome, and a thoughtful young man greeted us.

"Good evening, Sirs. May I help you? I apologize, but we are quite busy just now, so if you'll be brief."

"Indeed so," said Holmes. "I am Sherlock Holmes, and this is my colleague Dr. Watson. I wonder if we may have a talk with the resident astronomer?" We were shown into a waiting room with various timepieces and photographs of stars and nebulae on the walls.

"I say Holmes, isn't this a rather peculiar time for a tour of the observatory?"

"What better time to visit astronomers than at night?" he replied, "But I see here is our host."

A small, brisk older man entered the room. He had on large glasses, an overcoat, and was holding several lenses in his left hand. "My goodness, Patrick, we are going to miss the primary eclipse of Algol if we don't hurry! Can I help you gentleman?" he said distractedly.

"I apologize if my timing is not in line with the stars, Sir Barrington," said Holmes, "but if I could have a word with you for five minutes, it could be of great importance."

"All right. Five minutes then...if you'll follow me."

"Watson, perhaps you and Dr...."

"Patrick is just fine..."

"Perhaps Patrick can show you around for a few minutes."

We looked at the pictures and I asked about the art of astronomical photography. In five minutes, Holmes emerged with Sir Barrington, who had a grin on his face. "I must say, your Mr. Holmes here has some unique applications for our equipment," he said. As we left, I must say that I have rarely been as perplexed at the thought processes of my friend as I was at that night. But he was deep in thought on the ride home and, as always, I thought it best not to disturb him.

When I came down for breakfast the next morning, my colleague was already up. It could have been that he had stayed the night in his corner chair with his pipe well stuffed. However, he had a square look of satisfaction on his face and held up the morning paper to me. "Mrs. Hudson, could we have some breakfast for Dr. Watson? Look at this."

The paper was folded back to an article about a gold exchange to take place at the Westminster Bank. "Holmes, do you suspect the vanishing robbers to try for that gold shipment?"

"Indeed not, Watson. But I believe they will try for the currency placed in the bank for the exchange. Yes, I believe that is where they will next strike."

"Ah, but when Holmes? Tonight?"

"I don't know. It is not very pleasant out. A bit damp and foggy. It may even rain tonight. I think we can relax for now. By the way, have you ever seen my monogram on the various types of fibers – hair, grass, wood, and so on – that can be used to identify the material and sometimes the location of the objects that criminals have been wearing or carrying at the time of a crime? Most interesting."

"Like the straw reeds we found at the Bank of London?"

"Indeed, Watson. They come from a certain kind of very strong straw grown in only one region of France and very uniquely applied. A most interesting study." We chatted on then about the weather and a bit about horse racing.

As the day passed on, I was somewhat involved with several of my patients. Holmes turned to his chemical analysis of soil salts. It rained again that day, but the sky was clear again when we met for dinner. "Beautiful evening, eh, Watson? Cool, clear. I think we'll have a robbery tonight, if the barometer is not lying."

"Surely the barometer cannot effect the criminal inclinations of men, Holmes?" I replied, remembering my renewed exasperation at this ongoing mystery.

"Indeed, Watson. But we shall know soon enough."

Calling Billy up to our room, Holmes sent two telegrams, one to Lestrade and the other to Sir Barrington at the observatory. Most peculiar, I thought of Holmes sudden interest in astronomy after dismissing it to me previously. Yes, most peculiar, thought I.

Within the hour a telegram came. "This is it Watson, the telegram from the observatory. We have a robbery going on. Bring your pistol."

We grabbed our coats and were off. "To the Westminster Bank!" I heard Holmes tell the cab.

We were off through the night at a rush. I must say that I was quite perplexed at what the observatory might have to do with the bank robbery, but we soon arrived to find an excited Lestrade in front of the bank. "They have escaped, Mr. Holmes. And most peculiar, too."

"Have the firecrackers gone off yet?" Holmes asked.

"Why yes!" said Lestrade, "We heard the firecrackers going off same as before, just as we arrived. We picked this one up setting them off."

We saw a small, snarling man being held by the nape of his collar by one of the larger constables. He sneered at us and said "Ah, so you think you have me. But I ain't done nothing. What, setting off a few 'crackers ain't no crime, you know."

"Disturbing the peace is," said Lestrade.

"So is accomplice to a robbery," said Holmes. "We know how it was done. Now you can make it easier on yourself if you tell us where your two accomplices have gone. We shall find it out one way or another."

"You ain't got no squealer here," said the small man, and began to struggle.

"All right, take him away, if you are finished with him Mr. Holmes," said Lestrade and my friend nodded. "Now what is this, Holmes? We find the same footprints, two of them going into the bank, and none coming out. We haven't found any tunnel, but we find three high-pressure bottles again. But this time they were left in a hurry on the floor."

"Ah, then they suspect we are on to them now. We had better get them this time."

"But they have gotten clean away."

"Not yet, Lestrade. But hold it...this may be just what I'm waiting for."

A young lad rode up on a bicycle. "Telegram for Mr. Holmes!"

It was from the observatory. Holmes stuck his finger in his mouth and held it up. "Yes. The wind is certainly right. Gentleman, to the meadow just off Tilbury Road! And let us hurry!"

"Holmes," said Lestrade, "you were right about the Westminster Bank, but why are we rushing off to a pasture outside of town?"

"Because, Lestrade, it is where we will find our bank robbers," Holmes said, rather enjoying our perplexity, I think. And we were off at a gallop through the night.

In a few minutes, we arrived at a pasture just off the road in time to see an amazing sight. As our carriage slowed, there were two men in a basket being pulled along beside us, apparently coming from nowhere. They were down in the large basket and didn't see us hopping out to intercept them. The basket scraped along the ground until it tipped, and the men and three large bags tumbled out. It was then that I saw that ropes were attached to the basket and were tethered to a great balloon apparently filled with gas or hot air. Quickly, our guns drawn, and taking them completely by surprise, we rushed up to them and told them to stand up.

"How in the world...why even we don't know where we are going to be landing," said one of the men with a look of amazement on his face. Getting up out of the mud, we could see that these two men were rather slightly built, most appropriate for their method of escape.

"Lestrade," said Holmes, "I think you'll find the Westminster Bank's currency in those bags. It should not be difficult to obtain the other funds soon, as well."

"You sure do owe us an explanation on this one Holmes," said Lestrade handcuffing the prisoners. A constable led them away.

"I can now understand the bottles of gas being found. They hid them so that we wouldn't suspect their means of escape. But how about the firecrackers?"

"Why, doesn't a balloon make a lot of noise when it is first being filled with hot air?" I said. "I heard them once at a fair where the bottles of gas made an awful roar when heating up the air inside the balloon."

"Exactly, Watson," said Holmes. "The third man was covering up the noise on the roof with a firecracker barrage. Of course you now see that the men were carrying those high-pressure bottle-tanks in their balloon basket, the latter the source of the peculiar crisscross pattern in the mud and the source of the little straw reeds. When I found out that balloon baskets are only manufactured in a certain region of France, and only of this unique type of straw plant that comes from there, my suspicions were confirmed."

"But how did you know when the second robbery would take place, Holmes?" I asked.

"They certainly needed clear weather to fly away, but the cold air also helped them to get airborne. I told Sir Barrington at the observatory of my problem in tracking a flying object over London and asked if the stars couldn't wait until we had spotted these two over the bank."

"Of course!" I said.

"Well, he didn't have much trouble picking them out in the bright Moon tonight. Their first exploit had made them rather bold. He spotted their balloon when it was inflated over the Westminster Bank and reported their position to me when they were coming down. Probably low of fuel, they had to come down very near where they were last spotted, and we overtook them in time to see them land."

"Well, if that don't beat all," said Lestrade. "We thank you again, Mr. Holmes," and he was off with the prisoners.

"We should be in time for a late supper," said Holmes as we entered the carriage and started back to Baker Street.

"A most singular method of getaway," I said, as Holmes wrote a telegram to the observatory. "It certainly added a new dimension to our investigations, eh, Holmes?"

"Assuredly," smiled Holmes. "I must say that while I've been noted for my use of the magnifying glass, until now the criminological applications of the telescope had eluded me. Yes, indeed, Watson, one could certainly say that business has been looking up."

About the Publisher

Moira Doyle, a retired instructor of fashion design at the Fashion Institute of Design and Merchandising, and founder of Red Thistle Publications, is the sister of Dr. Laurance Doyle, and a triplet along with their brother Eric. She spent many summer vacations on Mt. Hamilton, which houses the Lick Observatory, assisting Laurance with his astronomical observations, and is listed as a coauthor in two of Laurance's pioneering transit extrasolar planet detection methods in The Astrophysical Journal.

Watching Laurance conduct his research with no outside funding motivated her to start Red Thistle Publications as a way to share his adventures and discoveries with a wider audience. Her company took its name from Mt. Hamilton's thistle plants, which are unlike those Moira has seen elsewhere; instead of the typical purple, they're a beautiful scarlet red. Her wonderful memories of spending time on the mountain with Laurance inspired her to use these red thistles as the namesake of her company, and her thistle/observatory logo is a fitting tribute to both Mt. Hamilton and her brother's work.